KID
FRIENDLY
COMPUTATION

ADDITION & SUBTRACTION

SARAH MORGAN MAJOR

Zephyr Press

Chicago

Addition and Subtraction

Grades: PreK–3

© 2002 by Sarah Morgan Major
Printed in the United States of America

ISBN: 1-56976-199-X

Cover: Rattray Design
Interior Design: Dan Miedaner
Illustrations: Sarah Morgan Major

Published by: Zephyr Press
An imprint of Chicago Review Press
814 North Franklin Street
Chicago, Illinois 60610
(800)232-2187
www.zephyrpress.com

 Zephyr Press is a registered trademark of Chicago Review Press, Inc.

Library of Congress Cataloging-in-Publication Data

Major, Sarah Morgan, 1953–
    Addition and subtraction / Sarah Morgan Major.
       p. cm. — (Kid-friendly computation)
    Includes bibliographical references and index.
    ISBN 1-56976-199-X
    1. Subtraction—Study and teaching (Elementary) 2. Addition—Study and teaching
       (Elementary) 3. Arithmetic—Study and teaching (Elementary) I. Title

QA135.6.M35 2005
513.2'11—dc22   2005005189

*With love always to Alex, Mitch, Beau, Emily,
and Adrianna, who taught me so much!*

Contents

How to Use This Book

You may approach this book in several ways, depending upon your particular needs, the level and ages of the children you are teaching, and your time constraints. If you have the time, reading the entire book is best. It is not difficult reading! Although the method of visual computing is not presented until part II (chapter 4), the chapters in part I contain essential background information. The chart on the following page shows the contents of the book followed by suggestions for use, depending on whether the children you teach are at a beginning or intermediate level, or are older children needing remedial assistance.

PART I: THE FRAMEWORK	
Chapter	**Content**
The Children	
Chapter 1: *The Whole Picture*	**Who is targeted and why:** Identifying the children who have difficulty with math and how to apply the *Kid-Friendly Computation* method with children of various ages.
The Practice	
Chapter 2: *Good Practice*	**Necessary elements:** Elements of good practice for successfully teaching math to all students.
Chapter 3: *Assessment Practices That Work*	**How to assess:** Using assessments as diagnostics and as tracking tools.
PART II: THE METHOD	
Chapter 4: *Learning Numbers*	**Numbers:** Activities and lessons for number recognition, counting, writing numbers, and ordering numbers.
Chapter 5: *Understanding the Meaning of Numbers*	**Meaning of numbers:** Activities and lessons to teach what each number means and how numbers are related.
Chapter 6: *Making the Transition from Visual to Symbolic*	**From visual to symbolic:** Activities and lessons for real-world connections, story problems, the action of computation.
Chapter 7: *Mastering Computation to Ten*	**Computing 1–10:** Activities and lessons to teach addition and subtraction with these numbers.
APPENDIXES	
Appendix A: *Reproducible Blackline Masters*	**Worksheets and templates:** All materials necessary to implement the lessons in the book.
Appendix B: *Tracking Charts and Progress Reports*	**Monitoring forms:** Reproducible forms for tracking individual student and whole-class progress, geared to the content of each chapter.

Introduction

Three Motivators to Write This Book

▶ Personal Experience

▶ Personal Research

▶ Personal Friends

My early love affair with math . . .

Addition and Subtraction is about teaching computation in away that accommodates the children who do not learn best with traditional math teaching methods. The groundwork for this method was laid throughout a lifetime of personal learning and observation, and what began as a significant interest has recently crescendoed into a consuming passion.

Personal Experience

I am the ideal candidate to write a book on teaching math. I did, after all, religiously fail arithmetic throughout elementary school. To this day, I can still see corrected papers piled in front of me on my desk, slashing red marks punctuating them liberally. In my imagination, I am once again a child slouching in my chair, while years and years of papers are heaped in front of me, each new one as disfigured as the last, until I am completely buried in my failure.

To me, the worst part about doing arithmetic back in those days was that I had to work on each returned paper until I got every problem correct. Sometimes this took so long that I was completely numb and eraser holes were worn clear through my paper. Nothing ever seemed to help me, and my inability to retain what I "learned" continued to plague me. We teachers tend to say about children like I was, "Oh, she's just not good in math."

In high school and college I did manage to get A's and B's in math, but the inability to remember stayed with me. Each time I was faced with problems to do, I had to figure out all over again how to do them, and as soon as the chapter tests were over, the information that had been fleetingly stored in my memory disappeared.

When I entered graduate school to study education, I failed the entrance math test and, completely panicked, ran to my daughter for help. I did pass the test the second time, but when it was time to enroll in my teaching of math courses, my anxiety knew no limits. When the professor gave oral directions, my stomach clenched. I could hear the words, I understood each word separately, but could not process them together and derive meaning out of them. In short, I didn't have a clue what I was supposed to do!

What finally turned me around in my own experience with math was that during my studies for other education classes, I recognized at long last that I simply could not retain what I learned in math when I was taught in the traditional way. As I spent valuable time discovering my learning preferences, I took the initiative to learn and "do" math in ways that suited my own learning needs. The by-product for me was a newfound enjoyment of computation!

Personal Research

Because I am not a traditional learner, I was drawn to books that discuss the different ways people learn, understand, remember, and are gifted. I felt as if I'd stumbled upon grace! Although I'd long since learned to compensate for what was lacking in my school experiences, it was wonderful to find out there was not a deficiency in my way of seeing and remembering. I felt validated. I have learned to embrace the strengths inherent in my particular way of processing and using information.

It soon became apparent to me that too often teaching practices are based on only a few of the many learning preferences, and that children who do not happen to fall into this narrow range often suffer tremendously in their learning experiences. Often these children are labeled with learning disabilities because they are unable to cope with the expectations in place within their school systems. Even at an early age, children can learn their particular learning needs and can acquire compensatory habits that will strengthen their school experience. But they must have an adult to guide them into this understanding early in their educational careers. The purpose of this book is not to teach learning styles, but where relevant I will refer to wonderful reading material that will open doors of understanding for teachers and parents who want to better understand how to reach all their students and help them achieve success with math.

Personal Friends

A few years ago, a fourth grader, Lisa, spent the weekend at my home. I asked her how school was going (one of those inane questions adults seem compelled to ask children). She remarked that she just couldn't learn her "times fours." Lisa had been in a resource room for two years already and was so defeated that she rarely communicated without saying things such as, "Oh, I don't know what I'm trying to say," or "I just can't say it."

My heart went out to her, so in the few minutes we had before dinner, I decided to try to find a way to help her learn her times four table. Out of sheer blind luck, I hit on something that made sense to her, and within 20 minutes, Lisa knew her fours. She knew them an hour later, and still knew them the next day. Two weeks later, when I saw her again, she still remembered them.

What struck me during this experience with Lisa was that there was obviously nothing wrong with her brain! There was not "something missing" in her that prevented her from learning. What was missing was an approach that was compatible with the way she and many other children learn best. When she was taught in a way that dovetailed with her way of understanding and remembering, she learned very rapidly.

The following spring, Lisa was tested for learning styles, and the test results showed that she is a primarily visual learner. No learning disability emerged during testing. Her school, however, did not "have a program to accommodate her," so when she entered school that fall, she was placed in a classroom for children with learning disabilities. The glass-walled classroom, located between the two regular classrooms of the same grade level, was dubbed "The Fishbowl" by the educators in that school. Lisa had to enter that glass room every morning amidst the stares of her former friends. At night she cried. The destruction of her confidence was complete.

Since that experience with Lisa, I have worked both in introducing math for the first time to the very young and in remediation with older children, usually third graders, who were failing in school. As I approached each child determined to discover his or her particular "best learning experience" and tailor my practice to those individual needs, I experienced tremendous growth. Over time, certain ideas emerged that seemed to be useful to almost all the children, regardless of their learning preferences.

I hope that you as parents and teachers, through using this book, will be filled with exuberance over the marvelous ways children's minds work, and that you will begin a lifelong habit of tailoring your teaching practice to individual children's learning needs. Although this method is excellent in remedial situations, I also hope that all teachers will begin to practice a method of teaching computation skills that will embrace all children's various ways of understanding, so that all of them can succeed within their regular classrooms.

Part I
The Framework

It was class as usual in Mrs. Swift's room!

Chapter 1
The Whole Picture

Mrs. Swift's class listens quietly to her directions

As I look back over the process of my schooling in how children learn, I see a road stretching out behind me—a winding road with bumps and hills. At every significant point, a child's face shines and I say to myself, "I learned that strategy from Lisa, this from Alice, this one from Ben, this from Nathan, that from Debbie," and the list goes on. I started to learn to teach when the children started to fill my heart and my vision. As I taught, I learned from the children how uniquely they view the world and process new ideas. The more I worked with and learned about them, the more certain elements of practice surfaced that seemed to work nearly universally, elements not found in traditional classes.

Considering Learning Styles When Teaching Math

To understand the value of this method of teaching computation, it is important to take the time to look globally at learning styles and how they relate to traditional ways of teaching math. This overview will lead naturally into identifying those learning needs that must be addressed in order for a math method to be successful for all children, regardless of their learning needs.

The whole topic of teaching to all those learning styles can be very intimidating to overworked teachers and parents. Because this is true, I first provide an orientation to the learning needs children have, and then pull all the ideas together and draw some conclusions that will help to bring the seemingly disparate elements together into a simple plan for good teaching practice.

The chart on page 4 presents an overview of the learning styles that I consider critical to this discussion. (For in-depth information on these and other learning styles, please refer to Barbe 1985; Gardner 1993; Gregorc 1982; Tobias 1994; and Witkin 1977.) If we imagine that the learning styles on the right side of the chart represent real children in real classrooms, it will become easier to see which children are being "taught around" in traditional methods of teaching.

Learning Styles in Traditional Methods of Teaching Math

Math is normally taught in tiny steps; students are given seemingly unrelated bits of information to work with or are given steps to memorize for solving problems. Often there is no real-life application within the problems, and all too often, students work solely with paper and pencil, having no opportunity to construct meaning for themselves using real objects.

Is it possible to teach math, as seemingly concrete and sequential as it is, in a way that will reach the abstract, random, visual/spatial, kinesthetic, and global students? Or should we continue trying to force them over to the left side of the chart? It seems more reasonable to change our current practice to fit the children rather than trying to force children to be something they cannot be. Let's make the assumption, then, that we should expand our method of teaching math to encompass and embrace all our students. What we will do in this book is approach computation in a global, visual, kinesthetic, abstract, and random way so that no child is left out!

Traditional Methods

Love These Kids!	Leave Out These Kids!

How Do I Learn? (Dr. Anthony F. Gregorc 1982)

Concrete
I use my senses to take in data about the world. What I see is what I get.

 I perceive the world

Abstract
I visualize, intuit, imagine, read between the lines, and make connections. I pick up subtle clues.

Sequential
I organize my thoughts in a linear, step-by-step manner. I prefer to follow the plan.

I order the information

Random
I organize my thoughts in segments. I will probably skip details and even whole steps, but I will still reach the goal. I like to make up my own steps.

How Do I Remember? (Raymond Swassing and Walter Barbe 1999)

Auditory
I listen to directions. I need to hear the sounds.

Visual
I need to see it. I make visual associations, mental maps or pictures, and see patterns.

Kinesthetic
I remember well what I learn through my body. I learn best by actually doing the job.

How Do I Understand? (Herman Witkin 1977)

Analytic
I am good with details, can follow steps and hear instructions, and like to finish one thing at a time.

 1, 2, 3

Global
Show me the big picture! I need to see how all the parts fit in. I can hear directions after you show me the goal.

How Am I Smart? (Howard Gardner 1993)

Verbal/Linguistic
I am verbal! I can speak, write, debate, and express myself well through words. IQ tests love me!

Visual/Spatial
Show me a map and I'll have it! I make vivid mental images and can use these to recall associated information. I want to see how something fits into its environment or surroundings.

Logical/Mathematical
I rely heavily on my logic and reasoning to work through problems. I am a whiz on standardized tests!

Body/Kinesthetic
I combine thinking with movement. I do well with activities that require precise motions. I learn by doing; my attention follows my movements.

4

The Common Denominator

I have come to believe that children who are highly visual also tend to be global, somewhat random, and kinesthetic. Think about it. Visual children see a whole picture, see smaller elements within their environment, see their connection to other elements in the whole picture, and tend to remember parts of the picture based on where each part fits into the whole. In addition, highly visual children will move randomly through the picture (or map or pattern) and are often inclined to spatial activities that require physical skill. Visual children will prefer to see the task done as they learn it, rather than hearing it explained, and will profit from doing the problem themselves. They might not understand the process the way another student sees it, but if they are certain of the goal of the lesson, they will likely invent good steps that make sense to them and allow them to reach the goal.

Learning Disabilities?

By now it might be apparent that I prefer to avoid labeling children in any way. Ever since my experience with Lisa and the fishbowl, I've been trying to learn as much as I can about learning disabilities. What are they really? Poor eyesight or hearing obviously qualify, as they could hinder learning if left undetected. Often, however, when I have inquired about the nature of a particular child's disability, a get a vague answer such as, "She has a reading disability," or "He just cannot remember." My translation for the first might be, "Someone never taught this child to read." For the second I would be tempted to think, "No one has discovered this child's learning style." At other times, I might conclude, "He has gotten into the habit of daydreaming," or "She is waiting to be told the answer and is not thinking for herself," or "Her role in life to date has been to be cute—someone needs to show her the value in problem solving."

Children can learn to exercise metacognition if they are guided in that direction by an observant adult. They can learn to compensate for differences they might have in learning, and they can improve their habits.

> ☆ ☆ ☆ CASE STUDY ☆ ☆ ☆
>
> I worked for months with a child who never had been expected to think for herself, or frankly, do anything more than be very cute. She expected me to tell her all the answers. I worked with her for some time on forming new personal habits for learning, and one day she said to me, "I'm blinking because my brain started to go to sleep, and I'm making it come back and think." She was only five.

Teaching Styles

In addition to my passion for enabling every child to learn, I have a passion for teaching in such a way that every child in the group can learn without being pulled out for "special help." It would be very difficult for overworked classroom teachers to present a lesson in seven or eight different ways in order to speak to the various intelligences identified by Howard Gardner (1993), and I am not sure doing so is necessary.

I have been growing into a teaching style that incorporates the three modalities (visual, auditory, and kinesthetic), thus encompassing three pathways to the brain. I then encourage the children to use their own strongest intelligences as often as possible when they do their projects. The approach becomes a flow: three pathways in, and several intelligences out. This practice has become my discipline and has carried with it a large reward: that of seeing children jump ahead in learning.

The auditory modality is our primary means of communication. Visual and kinesthetic, though, are immensely powerful allies. Put all three together and you experience magic. This math method is the result of my search for a way of teaching to these three modalities. I have used it for every child regardless of age, learning style, previous experience, or grade in school. What has resulted is that both the "quick learners" and the "slow learners" were able to succeed. (Quick learners are those who can make connections for themselves as they learn. Slow learners simply need some help making vital connections for learning.)

Teaching to Various Ages and Learning Styles

Through experience I have found certain generalizations about how children of different ages approach learning.

Preschool and Kindergarten

As a rule, for very young children, learning rate is relatively consistent. The younger the child, the slower the learning because the amount of prior knowledge is limited. Preschool and kindergarten students need a lot of practice in a very non-pressured environment with the child in full control of the pace. Children of five or six have had relatively little experience with numbers and will need to learn to extract meaning as they are exposed to these activities. Another feature of interest is that these young children will work away, constructing meaning as they go, then suddenly, several elements will click into place for them, and they will seemingly jump ahead in their ability to manipulate and use numbers.

☆ ☆ ☆ CASE STUDY ☆ ☆ ☆

My beginning group last year (eight children, ages four to six) started out slowly, using the activities in this book. Suddenly, around March, they mastered the contents of this book and moved well into *Place Value* which deals with computation using two-digit numbers. It was not my choice for them to go that far. I proceeded reluctantly at first, but in the end, I decided to follow where they led.

First and Second Grades

First and second graders who did not begin their math learning using this method may be doing well in school because they have learned to count on their fingers as they compute. With these students, the struggle you will face is that of breaking the habit of counting on fingers. Once they understand that you are teaching them a new way of doing math that involves seeing, rather than counting on fingers, they quickly adjust to the visual-patterned method and learn very quickly.

It is normal for children of this age to master computation for a specific target number—eights, for example—in one session. By this, I mean that they can do all sums to eight and numbers subtracted from eight. I have them spend the following week practicing that target number alone, then introduce mixed practice involving other target numbers.

Third and Fourth Grades

For nine-year-olds who have had a bad experience with math in third grade (for example, failing or being threatened with repeating third grade), the primary struggle will be to coax them out of the shutdown that occurs if they so much as catch a glimpse of a paper with double-digit subtraction problems. Because of their past failures, they may experience full-blown anxiety, glazing over and becoming unable to remember anything they have learned.

Many of these children have come to

expect failure so much that they are stunned and seemingly cannot think. The first hurdle with this group of children is to talk them down off the cliff to which they are clinging in their anxiety. When they finally understand that this way of learning math is nothing like what they have experienced in the past, they begin to focus on working toward mastery, and their ability to learn grows exponentially. But overcoming anxiety takes many positive experiences and successes. This summer, my nine-year-olds mastered computation to ten in a month. They did so with a one-hour weekly session with me and accompanying daily practice at home. Neither memorization nor counting on fingers was involved.

Traditional versus Nontraditional Learners

For those children who are able to learn well in a regular classroom, using the *Kid-Friendly Computation* method helps them learn to compute more easily, prevents them from developing bad habits, and takes the tedium out of doing math. For those very visual children at the opposite end of the spectrum, this approach gives them an equal opportunity to succeed by bringing the right and left sides of the learning styles chart together somewhere in the middle. Because this first book provides a foundation for computing with multi-digit numbers, children moving into larger numbers are challenged but not daunted. Their work in *Place Value* simply refines what they have learned in this book.

☆ ☆ ☆ CASE STUDY ☆ ☆ ☆

During my early association with Alice, I began to believe that I would never find a strategy that would enable her to learn (that is, make good connections in her learning on her own). I began to work with her so that she would have a foundation for the day when she would enter the regular classroom. Nothing I did seemed to stick with her. She spent more than a year trying to grasp what happens when you add two of something to one of something. I was stumped and, at times, frustrated. At five and a half, Alice still unconcernedly ate her way through the chocolate chips we used for computation (my attempts to add realism) and still did not retain anything. I battled within myself, wondering whether I should just give up on her and fall back on the practice of labeling Alice as a child who "cannot learn." I even had fleeting doubts about my whole philosophy that every child has the ability to learn—all because Alice did not remember 1 + 2.

My memory of the fishbowl kept me plodding away. I learned from Alice herself just how powerfully visual she was. She is the one who inspired the "my two hands" component (explained in chapter 6), which has proved to be a simple yet powerful visual-kinesthetic learning tool.

Alice happily digesting our math facts.

Five and no more.

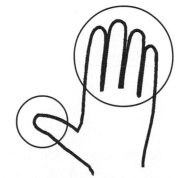

One and four is five.

Two and three is five.

Again, I hit upon this strategy out of desperation when I wanted to show Alice in one more way what was happening in the processes of adding and subtracting. This time, the approach clicked with her. Once we began to use "my two hands," Alice began to hum along. She took only four weeks to catch up with the rest of the class, who by this time had advanced confidently to the end of the tens.

TEACHING GUIDELINES

Now let's take these ideas and distill from them some basic elements of a good teaching approach that will include children from both sides of the learning styles chart:

1. State the goal first: "Today we are going to learn our facts for the number five, for example."

2. Provide concrete materials for the students to manipulate, establish clear but general parameters within which they will work, then let them discover the facts to five.

3. Communicate with the students about what they have discovered and guide them in drawing conclusions. This step involves pattern detection and exploration.

4. Use real-life examples of using these sums. Use stories whenever you possibly can.

5. Allow as much practice in solving problems as the students need.

6. Don't expect the students to "just remember" anything. Instead, tie every new concept to a previously learned concept, using visual and movement cues.

7. Develop a habit of teaching to all three modalities.

In the next chapter, I discuss some practices upon which this method is based, practices that will result in a good learning experience for every learning style. These practices relate directly to the teaching of math and will ensure that this visual method will work for you!

Chapter 2
Good Practice

Caleb telling Mrs. Swift he would rather be in the *Blue* group!

Introducing computation in a visual way is one element of good practice, but without the right environment, it is doubtful that a new method such as this one would be effective. This chapter will touch briefly on 13 principles that will maximize the success of this method. Creating an environment conducive to good math learning might mean making some adjustments or even learning some new tricks. But the payoffs are enormous.

When I followed the guidelines in this chapter with my own students, they made tremendous gains in terms of how much they absorbed, but best of all, not a single child dreaded math or avoided it. These students have moved on into "regular" classrooms, and each of them has stunned his or her teacher by asking for more math or more difficult math. In my classroom, these children were used to asking, "Can you give me some math to do?" They were used to hearing, "What kind of problems would you like today?" If this sounds like a fantasy, let me assure you it is not.

Mrs. Swift just moments after Alex asked for more math ...

Principle 1. Engage the Emotions

Guideline: Engaging the emotions enhances learning by creating positive biases toward math and may assist cognition.

Rationale

Much good information has been written on the subject of the interaction of emotions and cognition (see Hart 1999; Jensen 1998; LeDoux 1996). Increasing numbers of researchers are taking the position that emotions and higher-order thinking interact (see, for example, LeDoux 1996). The bottom line for our purposes, however, is simply that the more pleasant you make the process of doing math, the more the children will like math—that formerly hated and avoided subject. Once you start to look for ways to create positive biases toward math, you will think of many ideas that will work well in your own classroom.

Applications

To promote positive feelings about math in your students, try the following:

- Remember the senses: Let a subtle scent permeate the classroom.
- Light a table lamp or two— only during math time.
- Play classical music softly (no words to distract).
- Set a mood of anticipation, smile, and model positive feelings.
- Have special pencils reserved just for math use.
- Use pretty colors of paper now and then for variety.
- Let the students choose a fine-tipped marker in a favorite color for their practice.
- Choose visually appealing charts and posters depicting math patterns.
- Let the children illustrate their papers to show what is going on in the problem, or just let them decorate their work.
- Whenever possible, introduce the lesson with a short story that will engage the children and set a context for the lesson. The story might be about a situation that has arisen which the students will need to resolve.
- Approach the subject with the wonder it deserves.

Principle 2. Reduce Perceived Threats

Guideline: Negative experiences or threat during learning may detract from a student's ability to learn.

Rationale

When the brain perceives a threat in the environment, it initiates a "fight or flight" response, the familiar adrenaline rush. Not only physical threats but also emotional or environmental threats may trigger this reflexive response, and what is perceived as threatening varies from individual to individual. To say that threat "shuts down" the brain is an overstatement, but in a threatening situation, one's brain is at least partially occupied with evaluating the threat and planning possible responses to it. This leaves fewer mental resources for creative problem solving and learning. According to Leslie Hart (1999, 204), "Cerebral learning and threat conflict directly and completely." Hart identifies the following cognitive processes as potentially being disrupted by threat: pattern discrimination (which forms the backbone of this method), program selection (that is, "a fixed sequence for accomplishing some *intended objective*," such as solving a problem), the use of oral or written language, and symbol manipulation (Hart 1999, 154, 204). Obviously, if our goal is to create an environment conducive to learning, minimizing actual or potential threats in the learning environment is part of that effort.

Applications

Examples of possible threats include the following:

- Making a child do a problem on the board, then announcing that the answer is wrong

- Giving tests for which the child is not ready (resulting in poor grades)

- Ranking children in ability groups

- Forcing the whole class to progress relentlessly in lockstep despite some children's boredom and others' need to spend a little more time on a particular set of numbers

- A spirit of combativeness or competition in the group that results in the defeat of particular children, rather than a focus on whole-group cooperation and success

- An emphasis on speed and perfect papers rather than mastery, growth, and thinking processes

Principle 3. Facilitate Pattern Discovery

Guideline: A child more readily learns material if it is embedded within a pattern.

Rationale

Brain research tells us that the human brain is a pattern-seeking organ, and we have only to watch children at work in pattern discovery to know that this is true. Patterns provide order for seemingly chaotic material and help us make sense out of seemingly random facts. Some would even argue that recognizing patterns is an innate ability (Caine, Caine, and Crowell 1999). As Leslie Hart concludes, "Learning is the extraction, from confusion, of meaningful patterns. . . . Even rote learning is greatly helped by detecting the patterns involved" (Hart 1999, 127). Pattern discovery also provides the global framework so important to many children. The greatest benefit of pattern discovery is that it eliminates the need for memorization of facts. Through practice, students will come to remember the pattern visually, like having a mental snapshot stored in their minds, so they can locate individual facts from within that global picture.

Applications

Suggestions for encouraging pattern discovery include the following:

- Allow time for pattern discovery.
- Model your own habit of pattern discovery in everything you teach students. Ask, "How can we arrange the problems so that they form a pattern?"
- Give your students groups of facts at the same time so they can discover relationships between the problems, rather than giving them isolated problems with no apparent connection to each other (see box on page 16 for an example).
- Teach groups of facts that have something in common. (For example, present problems that have the same sum, sums that increase by one, or an addend that is the same.)
- Study and practice math facts within a pattern first, then mix them up, and finally, combine them with other, previously learned facts.

★ ★ ★ HOW TO FACILITATE PATTERN DISCOVERY ★ ★ ★

Consider the following seemingly random problem set:

$$\begin{array}{r} 2 \\ +1 \\ \hline 3 \end{array} \qquad \begin{array}{r} 4 \\ +2 \\ \hline 6 \end{array} \qquad \begin{array}{r} 3 \\ +2 \\ \hline 5 \end{array} \qquad \begin{array}{r} 5 \\ +4 \\ \hline 9 \end{array}$$

Instead, you can provide a set of problems with an inherent pattern, such as this:

$$\begin{array}{r} 0 \\ +6 \\ \hline 6 \end{array} \qquad \begin{array}{r} 3 \\ +3 \\ \hline 6 \end{array} \qquad \begin{array}{r} 4 \\ +2 \\ \hline 6 \end{array} \qquad \begin{array}{r} 1 \\ +5 \\ \hline 6 \end{array}$$

Encourage students to rearrange the problems in an order that reveals the pattern:

$$\begin{array}{r} 0 \\ +6 \\ \hline 6 \end{array} \qquad \begin{array}{r} 1 \\ +5 \\ \hline 6 \end{array} \qquad \begin{array}{r} 2 \\ +4 \\ \hline 6 \end{array} \qquad \begin{array}{r} 3 \\ +3 \\ \hline 6 \end{array}$$

Principle 4. Use a Constructivist Approach

Guideline: Children will better understand and remember what they have worked to discover.

Rationale

It makes sense, doesn't it, that what we work out on our own will stay in our memory longer than something someone simply tells us? The very process of working toward a specific goal will cause us to remember both what we tried that did not work and what we discovered that did work. Handing students a paper with the sums to five written on it will result in limited rote learning. In contrast, imagine the learning if we challenge students to find out which pairs of numbers will equal five, give them five counters and two bowls, and let them figure out the answers by trial and error.

Applications

Here are some suggestions for using a constructivist approach:

- Let students figure out the rules for themselves whenever possible. You are there to stimulate the learning process, not spoon-feed answers.
- Ask questions that promote "what and why" thinking, such as, "How many ways can we arrange these counters so that each new combination equals seven?" "Why do you suppose this number got bigger?" "What would happen to this problem if we changed the addend?"
- Give students real materials to represent numbers and encourage them also to draw pictures of what is happening in a problem.
- Provide general guidelines, such as instructing students to write down all the pairs of numbers that equal seven, then crossing off any repetitions.
- Provide specific goals for discovery.
- Design your lesson so that students will gain meaning for themselves. If you want them to discover a certain rule, choose only problems that exemplify that rule, so that in time the students will notice the rule for themselves. Make sure to alert the students that they are looking for a rule as they work.
- Instead of teaching shortcuts, give your students enough practice with similar problems so that eventually they will find shortcuts for themselves.

Principle 5. Give Immediate Feedback

Guideline: Learning is enhanced by including in instruction a tool for immediate feedback.

Rationale

A constructivist approach that utilizes pattern discovery will be greatly enhanced if some means of immediate feedback is integrated into it. Feedback is a critical ingredient in the recipe for success. The children must have a way of knowing whether the pattern they have discovered is accurate. If they find they are wrong, they must be able to make immediate revisions.

Applications

Feedback does not have to come from a busy teacher! A carefully planned activity should include some form of feedback inherent in the process. Built-in feedback is a powerful teaching tool that offers students a sense of control over the learning process: if they can determine right away that their solution is not correct, they can continue their learning by searching for another answer. Here are several ways to provide instant feedback:

- Limit the number of materials given to each student. For example, if the goal is to discover pairs of numbers that equal five, give the children each five counters and two bowls, and instruct them to use *all* the chips each time they find a pair of numbers. They will not include the pair 1 + 2 because it does not use all their chips.

- Provide clear guidelines to direct their work. (For example, "You must use all the counters each time," or "You may not repeat the same combination of numbers.")

- Print each problem on an index card and write the answer on the back. The student can work the problem, then turn over the card to check the answer. Laminated practice sheets with dry-erase markers are a wonderful way to apply this concept with reusable materials.

- Have students work in pairs, with one partner working the problems and the other having the answers. (They can take turns being the one with the answers.)

- Teach kinesthetic or visual feedback methods that children can use independently, such as "my two hands" (page 64).

Principle 6. Work for Mastery

Guideline: The goal of instruction is for every student to master the material.

Rationale

The goal in teaching math must be mastery rather than grade collection. Every child can learn math, and if making this happen is our passion, we will do whatever it takes to achieve that goal. Success breeds more success, and once the threat of failure is removed, children can focus on learning.

An obvious reason to work for mastery by every child is that this goal most closely matches the reason we are in a classroom in the first place. Teachers are not grade collectors or classifiers of children; we are there to teach so that our students will learn. I partner with my students in this journey toward mastery.

Applications

Here are some suggestions that will help you in building a community that works toward mastery:

- Tell the children the learning goals for each section. (For example, "Today we are going to learn all our facts to ten.")
- Assess student progress frequently. (Ideas for successful assessment are provided in chapter 3.)
- Provide reteaching or practice immediately after an assessment as needed to reinforce weak areas.
- Make a habit of giving brief, very frequent reviews.
- Help your students to discover their learning preferences and tap into those strengths.
- Stress to your students that they are able to learn the material.
- Emphasize hard work over "being smart."
- Set short-term goals in addition to the long-term goals.
- Enlist parent and volunteer support as often as possible for tracking progress and conducting additional practice with specific children.
- Use student pairs for tracking progress and for additional practice.
- Build a community spirit in your classroom, in which students become accustomed to helping each other learn.
- Develop a habit of finding new ways to teach a concept whenever your current method is not successful for all children.

Principle 7. Achieve Mastery through Practice

Guideline: The brain will become fluent in those activities where it receives ample practice.

Rationale

Practice in a nonthreatening environment is a critical element for successful learning. Rote memorization simply does not work for some children. If, after drilling with flash cards, some students still don't know their math facts, it is doubtful whether more of the same will ever produce the desired result. I say, "Give it up! Memorization is a waste of time!" Frequent practice is, after all, the way humans learn anything that is not genetically stamped on their brains. With sufficient practice, the process of deriving math facts becomes automatic and the child says, "That's easy!"

Applications

Some suggestions for replacing rote memorization include these:

- Do pattern discovery as discussed previously.

- Use a constructivist approach.

- Discover the method(s) that work best for each child.

- Target particular elements for learning, rather than teaching a broad range of skills simultaneously. Explain the goals to the students at the beginning of the lesson.

- Use informal assessment not for the purpose of collecting grades, but in order to isolate facts not yet mastered.

- Make sure to ask each student which facts he or she has not yet mastered.

- Give practice sheets (which may be laminated for repeated practice) and ask the children to go through the page answering only those problems they recognize on sight.

- Take the time to enrich learning. Ask such questions as, "What else can we discover about this problem? What is it like? How is it different from the one next to it?"

- Take the time to help students form additional connections for problems they answered incorrectly.

- Review missed problems using concrete materials, repeating the practice daily, if possible.

- Reassess to check for mastery.

At this point, the student will be ready to move on to the next area of mastery. Because the child is in charge of this learning process, it is safe to let the child think about what he or she needs to practice. If you let your students take the initiative in identifying what facts they need to practice, their overall progress will be more rapid than if you set the pace.

Principle 8. Provide Visual Connections

Guideline: Find a means of visually connecting every new concept to something the student already knows.

Rationale

Many scholars argue that most of what we remember enters the brain through the visual modality (Jensen 1994). When we make visual connections, either we are automatically reminded of something else, which is then linked in our memory with the new idea, or we make a conscious effort to form a connection that will serve as a memory prompt.

For visual learners, this avenue is essential. When visual learners hear verbal instructions, what they hear and what they are able to process are frequently two different things. They need to see what they are hearing.

Applications

Here are some suggestions for making connections as you teach:

- Ask students these questions constantly during your teaching: "What does this look like?" and "What does this remind you of?" Record students' answers.

- Discuss the various suggestions with the students and collectively agree on a specific visual cue for each concept. These cues become triggers that help students retrieve abstract facts from memory. Once the fact has been learned, children will automatically lose the need for the visual connection and recall will be automatic.

☆ ☆ ☆ CASE STUDY ☆ ☆ ☆

When my preschool students were having trouble remembering which number symbol corresponded to each number name, we discussed what each number looked like. The group chose known objects, the shapes of which reminded them of each number symbol.

Snowman, upside-down chair,   thin man,   unicycle

Continued

I watched and waited patiently during the early days of number learning, when the children used the number names and associated picture names interchangeably. Finally, the day came when they dropped the picture name and retained the number name. With the extra visual step of associating a picture, learning occurred more rapidly and without stress for the children. The children were easily engaged in learning their numbers. These stylized numbers will be presented in chapter 4.

Principle 9. Set the Stage for Visual Imprinting

Guideline: As often as possible, connect an abstract concept to a visual image that is meaningful to your students.

Rationale

Visual imprinting refers to a practice that is difficult to define. Imprinting just happens; it is a subconscious form of learning (what I call "learning through the back door"). Visual imprinting has occurred any time we can "see" a complete picture or a specific part of it in our mind's eye. Even though visual imprinting is elusive, we can deliberately take advantage of it in our teaching. In fact, it has become one of the most powerful tools in my tool belt. It is primarily through this means that my children learn their sight words, the meaning behind the number symbols, and their math facts to ten. Each time I ask, "How did you remember that?" and the student answers, "I saw the snowman" (see illustration below) or some other response that reveals visual imprinting, I feel the magic again. I have gotten goose bumps at times, have guffawed, and have even done my own middle-aged version of the touchdown dance.

Visual imprinting for the number 8.

Applications

To promote visual imprinting in your teaching, try these strategies:

- Use stylized materials as often as possible. For example, in helping a child remember 20, 30, 40, and so on, you can quickly make the 0 in each number into a teacup. Then, point to the first digit as you say the first part of the number name, and as you point to the 0/teacup in each number, say "tea." Let the pictures enter the mind of the child passively; do not try and actively teach the association. Remember, the child is acquiring a mental photo of the concept. This is an automatic process, not one that can be forced.

- Use such stylized materials only about three times during learning before stopping to check for recall using normal printed numbers. Isolate the facts the child still does not recall and use stylized materials with those facts a few more times.

- Recheck for knowledge using normal printed materials.

- Frequently ask various students, "How did you remember that?" This habit of yours will train the students to think about their own learning processes.

☆ ☆ ☆ CASE STUDIES OF VISUAL IMPRINTING ☆ ☆ ☆
Ethan

A fascinating example of visual imprinting occurred this spring as I was helping nine-year-old Ethan learn the multiplication table for eights. We were using an approach in which I provided a five-by-four grid of answers (see illustration on page 24). This global approach utilizes pattern discovery, visual mapping, and apparently also visual imprinting. After only 20 minutes, Ethan could do a sheet of mixed problems quickly and accurately without referring to the written answers and without counting. I was enthralled to see that as he worked, he would read a new problem, look at the empty grid, point to where the answer would be located, and then quickly write the answer. When I asked him what he was doing, he replied that he was finding the answers in the grid.

Continued

Because Ethan was concerned that he would not be able to recall the facts in his regular classroom, I asked whether he thought his teacher would mind if he taped an empty grid to his desktop. Horrified, Ethan said, "But that would be *cheating!*" His answer told me just how clearly Ethan was seeing the answers to the multiplication problems on the blank grid. I suspect that a combination of visual imprinting and visual mapping on a global pattern was occurring.

Becky

One day Becky, who was nine, was humming through her sums for six using the "Stony Brook Village" approach, in which each sum is represented inside a house on a particular numbered street (see chapter 7). After discovering the global pattern for sixes, she was using her new knowledge to complete a practice sheet of problems. I pointed to a problem she had just finished and asked, "How did you remember this one?" She instantly replied, "Oh, that is the last house on the right."

Neither drill nor memorization had taken place; learning was fast and painless. We set the stage with a story, did pattern discovery, practiced writing the pattern three times, then began some solo flights through sheets of problems. Becky was promptly successful.

Principle 10. Make Visual-Body Connections

Guideline: As often as possible, combine visual materials closely with body actions or movements.

Rationale

The visual modality is a powerful means of learning, but learning is enriched and deepened when we involve the body in the process. Doing so utilizes two powerful means of remembering simultaneously.

Applications

So, to strengthen your teaching practice, try these tips:

- Make connections between visual images and physical sensations, activities, or movements. For example, you could have a child make a dot card to represent the quantity three by dipping three fingertips into colored paint and then pressing them on an index card. The combination of pressing three fingers on the card and seeing the resulting dot pattern will result in significant visual-body connections as well as providing an image for imprinting (see illustration). In this activity, the child feels three fingertips getting wet and slippery and three fingertips pressing on the card. Beyond that, the child unconsciously gains a visual image of "threeness" not only by seeing the fingerprints on the card, but also by seeing in memory those three fingertips colored with paint. (This activity also leads naturally into the "my two hands" strategy, described on page 64.)

- Deliberately build triggers for recall in your students' minds: The difference in practice between doing this kind of activity versus simply counting three objects, or providing preprinted dot cards, is subtle but significant. After doing the activities, students will have images of number meanings lying stored in their memories. They will see three instead of counting to three. Similarly, the "my two hands" method of computation is a kinesthetic means of learning that also promotes heavy visual imprinting. Many children actually see mental images of the sums after having practiced them on their hands.

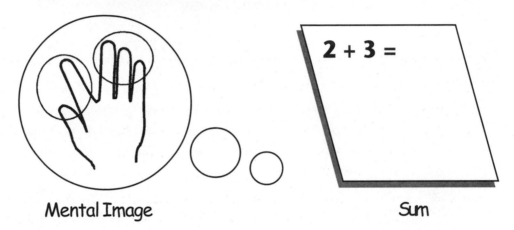

Mental Image   Sum

Principle 11. Make Associations and Connections

Guideline: Teach every new concept as an outgrowth of previous learning and a connection to future learning.

Rationale

Associations and connections are the best aids for memory and recall. When the brain is confronted with new, seemingly chaotic information, it gets busy trying to form patterns, discover connections to other knowledge, and make associations that will help it to make sense of and remember the new information. Thus, math learning can be greatly enhanced by these strategies:

- Assist the child to form associations with real events.
- Connect the steps in the learning process to one another.

Rather than teaching math in unrelated segments, we can start out purposefully with simple concepts upon which more complex skills will be built. This practice is consistent with research showing that the brain learns a basic skill and then develops more specialized versions of that initial skill (Hart 1999, 177). Start small, as you would when you pack a snowball in your hand. Then add to that first fact as you would add snow to the snowball by rolling it across the yard.

One example of how this strand runs throughout this method is the use of the number five. Students start by seeing what five looks like on their own hand (five fingers), then move to the use of a fives chart, then to using multiples of five as anchor numbers (that is, as reference points against which to describe the position of other numbers), to using groups of five in learning multiplication and division. All this is done purposefully to build on a simple concept that the child learned at the very beginning.

Applications

Here are some suggestions for making associations and connections in your teaching:

- Plan lessons that integrate new information with prior learning.
- Teach information in groups of related facts rather than as isolated facts.
- Find as many connections as possible to real-world situations.
- Tie new information to other disciplines.
- Ask questions to encourage the formation of associations; for example, "What does this look like? Remind you of? Sound like? How is it similar to . . . ? Different from . . . ?"
- Demonstrate how the new skill grows out of prior learning and will lead to future learning.

Principle 12. Associate Purposeful Movements with New Learning

---

Guideline: Include purposeful movement in the teaching of new concepts to forge a body-brain connection.

Rationale

It is magical to me how physical movements are stored in the cerebellum and become automatic. By automatic, I mean that they are not conscious or deliberate—the body just knows how to move. Think of riding a bike. Once we have learned how, we never forget. In climbing stairs, we know just how to slant our body so we don't fall backwards, just how far to raise our foot for the next step—which is why we trip when risers are not built to standard height.

If you are a musician, think of how you memorize a piece of music. At first, you read the notes carefully and may practice the difficult passages in isolation. After much practice, the body begins to remember. At this point, playing a song is a kinesthetic-cerebellar function, not a cerebral function.

This incredible ability of the body to remember should be utilized by embedding purposeful movement in learning as often as possible. By purposeful movement, I am referring to movement that directly relates to the concept learned, not fidgety or random movements like bouncing on a big ball or hopping during learning. Forming a number shape with the body is one example of a purposeful movement, and several more are suggested in the "Applications" section. What the body does during learning can be permanently embedded in the cerebellum, serving as a powerful avenue for learning and recall.

Applications

Here are some suggestions for incorporating purposeful movement into good teaching practice.

- Use hand motions, such as "my two hands (see page 64)," as often as possible during learning. Design hand motions that resemble the printed shape of an abstract concept (such as numerals or math operators) as often as possible.

- When learning a pattern, pair it with a patterned movement. For example, when counting by twos, the class can march around, leaning heavily to the right on each even number.

- Use movements for recalling the meaning of + and –. When you say "plus," bring your crossed arms to your chest, demonstrating that more are coming to you, mimicking the plus sign with your crossed arms. Similarly, when you say "minus," slash horizontally away from your body with your right arm, mimicking the horizontal shape of the sign as well as showing that the number is leaving you.

- Use full-body skywriting rather than writing in the air. Have students stretch their whole body into the shape of the number being written (see illustration). For best results with this type of skywriting, form a large number on the floor with masking tape. Have a child stand and try to duplicate the shape with the body, while you provide support so that he or she does not fall. Then have the child quickly write on paper what he or she felt in the body. The action of moving the body to form the number mirrors on a larger scale the movement the children will do when writing the number. Being able to see the number while body-tracing it integrates the visual and kinesthetic modalities.

- Have pairs of students make full-body numbers.

- Have students walk along a large masking-tape number on the floor.

- Engage the children in thinking of motions that remind them of the concept being learned.
- As often as possible, learn by the "see, say, and do" method. Give the children a visual image, let them say the concept, and get them to move in a meaningful way all at the same time.

☆ ☆ ☆ CASE STUDY ☆ ☆ ☆

During my early years of working with children, I did much of what I did because I thought it was "normal practice." I cannot say that I had a good reason for what I did. Take, for example, the daily routine of counting to 20 while one child pointed to each number in turn. It sounds mind-numbing now, but back then it seemed the thing to do, and the children didn't seem to mind the routine.

One morning, Peter's mother asked me if I'd taught the children (ages four and five) to count by fives. When she had tucked Peter into bed the night before, he suddenly began counting by fives. After a great deal of thought, I finally understood how he had learned this. When it was Peter's turn to point to the numbers, he heard the numbers, said them, and pointed to them—but beyond that, each time his finger reached a new "five" number, his body would turn to the left in order to start another row of numbers. "Through the back door," Peter learned to count by fives. I think his body taught him this skill by emphasizing the numbers he said each time he turned to the left.

Principle 13. Teach from a Global Perspective

Guideline: Teach new material from within a global whole and always specify goals for learning at the beginning.

Rationale

I have touched on the global nature of my approach. Eric Jensen, in his book *Teaching with the Brain in Mind*, recommends starting with the global view before working through the sequential steps. An analogy would be telling your class, "We are going to Chicago. Now here are the step-by-step directions." Not only is a global approach essential for those learners who simply must know where they are headed, but it is essential for pattern discovery. One cannot discover patterns without seeing the whole. Another benefit of the global approach is that it lets the students know exactly how much they need to learn. The list of facts does not seem to stretch out endlessly, in an interminable string that can never be learned. Instead, the students know, "This is all, and there is no more." The global learner needs this information.

Applications

Here are some suggestions for teaching from a global stance:

- Show the goal (such as learning numbers to ten) before starting the lesson.

- Show the children the whole of what they will be learning (such as Stony Brook Village— illustrated at right and described in chapter 7—or at least the whole street of numbers they will be learning).

Stony Brook Village

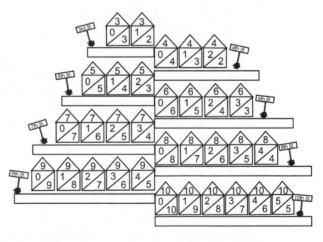

- Identify how each problem fits within the global whole, such as "On Tenth Street, 0 + 10 is the first house on the left, while 5 + 5 is the last house on the right" (see illustration).

- Encourage students to explore how each problem relates to problems on either side, noting similarities and differences. Brainstorm possible reasons for these similarities and differences.

☆ ☆ ☆ CASE STUDY ☆ ☆ ☆

I recall working with my group on math one morning. I had carefully chosen individual children's work based on where they were in their learning process. As I distributed the work, I noticed that Ben's face froze. He had taken one look at his paper and had a little anxiety attack. Casually I drew him aside to discover the source of his fears. He was able to verbalize that there were so many problems, and he was sure he didn't know them all. I knew something he didn't know, however, namely that every single problem on the page was a variation of these three number families: $1 + 6 = 7$, $2 + 5 = 7$, and $3 + 4 = 7$.

Instead of telling him this, I gave him seven counters and two bowls. I asked him to find out how many ways he could distribute *all* the counters into the two bowls without repeating any combination. He got right to work, assuring me confidently that there were "lots" of ways. He quickly found the three families, then began to falter, although still insisting that he would find more ways. I let him gnaw on that particular bone for a while. When he finally gave up, I asked him then to look at the problems on the practice sheet. Because of his brief discovery time, he quickly recognized that indeed he did know all the problems and that they were just variations of each other. Fully at ease, Ben confidently completed his practice sheet. What made the difference? The "all and no more" approach. All of these problems are derived from those three families:

$1 + 6$	$7 - 3$	$2 + 5$	$7 - 1$	$3 + 4$	$7 - 4$
$6 + 1$	$7 - 6$	$4 + 3$	$7 - 2$	$7 - 5$	$5 + 2$

Conclusion: Teach to the Children's Brains

In a nutshell, everything we know about how a child learns may be discovered by observing children as we work with them in the classroom day in and day out. If an approach is not working, of course we change it. If a child experiences distress or shuts down, we stop what we are doing. If one method is not working, we create another avenue to understanding. The bottom line is how determined we are that all of our students will learn, remember, and grow into mastery and confidence. This goal should matter more than anything. There is absolutely nothing like giving a child's brain what it likes! The traditional method of teaching children is rather like pushing a heavy lawnmower up a steep incline, whereas a method that is compatible with children's brains is like watching a robot-driven mower do the work while lazing on the porch sipping lemonade.

Chapter 3
Assessment Practices That Work

Mrs. Swift finally knows them all!

In order for the *More Math, Please!* method to work for you, a new attitude toward assessments must be combined with the elements of good practice outlined in chapter 2. Underlying everything we do is a deep respect for children that prevents us from engaging in practices that will damage or discourage even a single one. Respecting children means assuming the responsibility for taking them to the next level in their learning. This attitude is foundational to success in teaching all children. Without this attitude, it is possible to use methods that don't help the child, then to blame the child when learning does not occur.

Respecting our students and assuming responsibility for their learning causes a significant shift in our view and use of assessments. Rather than being an end in themselves, assessments become a tool to aid us in our quest for learning for all our students. The ideas in this chapter will help you successfully use assessments to support learning.

Informal Assessments

The heart, the passion, drives all else. If you have the passion, become a kid watcher. Notice what engages the children and what shuts them down:

1. Develop the habit of making notes for yourself as you observe your students at work. A clipboard or a notebook with a section for each student works well for this purpose. Several tracking forms are provided in Appendix B.

2. Target a handful of students to observe each day and record significant information regarding their work habits, learning methods, areas of mastery, and areas in which they are struggling. Here are examples of notes you might make:

    - Toni compares problem to problem as she works.
    - Juan uses "my two hands" as a strategy.
    - Lateisha closes her eyes and draws on an imprinted image.
    - Paul is struggling with 3 + 4 and 3 + 5.
    - Alicia is hurrying too fast through her work to read the problems correctly.
    - Lynn is making connections between this set of problems and the previous set we learned.
    - Takeshi struggles to discover a pattern even after many repetitions.
    - Eric looked back at a similar problem he'd already done.

Formal Assessments

Here are some suggestions for using written assessments.

Evaluate the Usefulness of Timed Tests

For the life of me, I cannot imagine a time in any child's future when being able to compute rapidly will save him from some catastrophe. What is the purpose of timing our students? Granted, there are some children who rise to the challenge of a timed test. Those children grow into adults who pay a lot of money to jump out of a tiny airplane wearing a parachute or leap off a bridge with a slender rubber band to prevent them from plunging to certain death. These people face challenges with an adrenaline surge and excitement.

For other children, timed tests represent a threat that interferes with their being able to think. The result of timing these children is that they freeze up, do poorly, and eventually experience tremendous anxiety around test taking. There are still other children who are wired in such a way that they simply cannot hurry. They are methodical, thorough workers who love to check their progress as they go along. They answer one question, adjust their body in the seat, look back at the well-answered problem, deliberately focus on the next problem, and repeat the pattern. Trying to make this type of child hurry will only result in frustration for the teacher, not in increased speed from the child. These are the students who may complete only the first half of a test, but the problems they complete are worked perfectly. Yet they receive a failing grade. Does this grade reflect their knowledge or their ability to compute? No. All it reflects is their speed, which would be much better determined in a footrace.

☆ ☆ ☆ CASE STUDY ☆ ☆ ☆

One day, I watched my children doing a sheet of double-digit subtraction problems. I watched particularly closely the progress of two boys who sat next to each other. One, barely five years old, quickly scoped out the nature of the work and literally looked and wrote his way down the page. He worked with energy and enjoyment. He "rabbited" his way through the page with amazing speed.

The other boy, age six, also had an air of being at ease. He worked the first problem, wrote the answer in his best handwriting, looked at his pencil point, looked back at the first problem, looked up and smiled at me, wiped imaginary eraser crumbs off his paper, then turned deliberately to the second problem. He "turtled" his way through the paper. He was not wasting time, nor did he daydream. He was being thorough and was proud of his work.

When I checked both papers, my rabbit missed five problems, which he easily corrected, while my turtle got a perfect score. Both knew how to do the computation. Both children are beautifully made. Both are smart. Both felt competent. But they worked very differently. Some people are not meant to be rushed.

Avoid Assessing the Whole Class Simultaneously on the Same Material

In order to administer fair formal assessments that accurately reflect each child's ability and knowledge, all the children would have to be at the exact same place in their learning processes. This is not a reasonable expectation. Remember Alice, who took months to master 1 + 2, then only four weeks (once I found her language) to master all the rest of the problems to ten? If I had administered formal assessments while Alice was in her period of discovery, she would have received failing scores. This failure would have communicated clearly to her that she could not do the work, taken her focus completely off her learning process, and diverted it to the process of test taking. Worse yet, her self-esteem would have suffered as she began to focus on comparing herself to the other students.

Design a System for Self-Pacing

Who is to say that a child must master computation to seven by the date set in the curriculum? Does it really matter when mastery is achieved? An important question to ask regarding such a schedule is, "Is this best practice for the child, or is this for adult convenience?" Again, if our goal is to educate, our measure of practice will be the child and his or her needs. Instead of testing all the children simultaneously, design storage cubbies in your room containing tests that the children can take as they are ready.

Give a Grade Based on Three Assessments

By averaging the results of three assessments, we can ensure that a particular score is an accurate reflection of knowledge, not a poor night's sleep or a bad day. Counting scores from three sessions also takes the pressure off the child and the focus off grades.

Use Assessments to Identify Areas Needing Practice

Use assessment results to identify children who need practice in the same areas. Group children with similar needs at math centers. Enlist parent or community volunteers to help lead practice sessions.

Record Progress and Mastery for Each Child

Progress charts geared to the method used in this book are provided in Appendix B. They are designed to yield information that is valuable both for you as a teacher and for sharing with parents. You will be able to tell at a glance where each child is in his or her learning.

Use the Progress Reports to Make Partners out of Parents

Focus on keeping parents informed of mastery, not grades. Work to create an atmosphere of being team members working together for the child's benefit.

Part II
The Method

Mrs. Swift using her very best tools

Chapter 4
Learning Numbers

Goals for This Chapter

1. To recognize numbers to ten (and to recognize the numerals that make up the Arabic number system)

2. To count to 20

3. To order numbers sequentially from one to ten (or 20, for those who are ready)

4. To supply missing numbers from within a global pattern

5. To discover patterns within a global whole

6. To write the numbers from one to ten correctly

Learning numbers goes far beyond learning to count and recognize the numbers on sight. In this chapter, we want to lay a rich base for later use of numbers. Consistent with our "see, say, and do" (visual, auditory, and kinesthetic) approach to learning, we will expose the children to numbers from several different directions at once. Students will get a visual image for each number, will see each number in context, will learn to write the numbers correctly right from the start, and will learn how to order them.

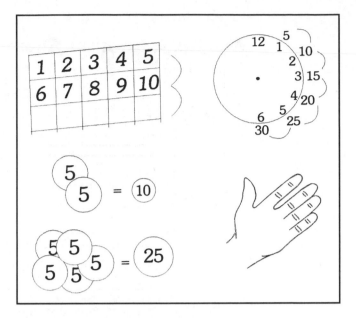

I arrange the numbers in rows of five from the very beginning. This arrangement is useful through basic computation and into multiplication and division. Consider all the applications of units of five: they relate directly to telling time in five-minute increments and to counting money. The set of five also corresponds to the number of fingers on the child's hand (see illustration). So we will start in the same way we intend to continue, by arranging the numbers in a format that will be used repeatedly later.

Next, we identify two rows of five. Ten is just as significant a group as five is. Ten becomes the next important anchor number. (An anchor number is a point of reference against which the location of other numbers is defined. For example, 12 is two more than ten.) There are ten fingers on two hands, and our number system is base ten, so this number relates directly to place value. We learn to compute to ten and use the exact method learned to that point for computation beyond ten. All of this is a deliberate attempt to connect the stages of learning in a smooth flow, where initial learning is small, easy, and general, and all new knowledge is added to this all-important beginning in the form of specialization (an application of principle 11 in chapter 2). So, let's get started! Here is what the first four rows of numbers look like:

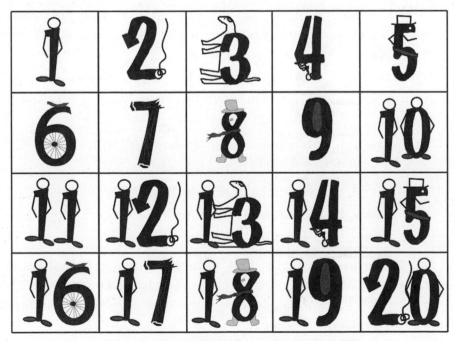

Chart of stylized numbers to 20

Making a Number Chart

Reproducible pages with the stylized numbers are provided in Appendix A. Worksheet 4.1 (page 88) is for individual student use or for number books and other student projects. For a group or classroom, make a large classroom wall chart, as described below:

1. Photocopy the pages of worksheet 4.2 (pages 89–93) actual size, or enlarge if you prefer.

2. Cut apart the numbers and paste in rows of five on a sheet of tagboard, as shown in the illustration on page 39.

3. If desired, laminate the chart or cover with clear plastic shelf paper for protection.

4. Position the chart prominently on a classroom wall so the children will view it often and visual imprinting will begin.

Materials Needed

- copies of stylized numbers (pages 89–93)
- photocopier
- scissors
- paste or glue stick
- sheet of tagboard at least 25" x 14"
- lamination equipment or clear plastic shelf paper (optional)
- place to display in the classroom

Using the Number Chart: Pattern Discovery

The children will make discoveries on their own if they have time to look at the chart and talk to each other about what they see, but some gentle guidance from you will jump-start the process.

1. Group the children around the chart. Ask them what they notice about it. They might point out that the same number appears more than once (for example, there is a 4 in 4 and 14), or they might just talk about the shapes of the numbers.

2. When the children have finished making comments, point to the 2 in the top row and ask, "Do you see another two?" (Point out the one two rows below.) Then discuss other, similar patterns:

   - The 1, 6, 1, 6 pattern in the left-hand column
   - The 2, 7, 2, 7 pattern in the second column
   - The 3, 8, 3, 8 pattern in the third column
   - The 4, 9, 4, 9 pattern in the fourth column
   - The 5, 0, 5, 0 pattern in the last column

3. Be creative in your exploration of these patterns to help the students make associations in multiple modalities. As often as possible, draw out discoveries by asking leading questions rather than just by telling the children what to look for.

4. Talk about similarities in the number shapes. For example, you might pose these questions:

"Do you think that maybe the 1 in the first row fell straight down and became curved at the bottom, forming the number 6?"

"Do you see a connection between the 2 and the 7?" (Both numbers contain a slant from top right to bottom left.)

"Look at the next column. How are the 3 and 8 alike?" (The 3 is just like an 8 with one side cut off.)

"Look at the 4, 9 pattern. How do the numbers look similar?" (Both numbers are top-heavy, with a line extending down to complete the number.)

"And look at the last column. What's similar about these numbers?" (Both numbers have a rounded part.)

If the number chart is extended, another pattern that will emerge is the 1, 1, 2, 2, 3, 3, 4, 4, etc., pattern in the tens place in each column.

Learn the Number Song

The music, lyrics, and suggested motions for the Number Song are printed on page 43. You may copy this page as needed. I have provided a very simple tune with guitar chords. If you have no musical training, you might ask the school music teacher or a musical friend of yours to make you a cassette tape of the music. Alternatively, sing the lyrics to the tune of "This Old Man."

The song is obviously a musical/rhythmic approach to learning numbers, and the lyrics provide a linguistic association to help in recognizing number shapes. The song, once learned, will assist your students in reaching four of the six goals for this chapter. It becomes a self-corrective strategy children can use as they learn to count in sequence (goal 2), find missing numbers (goal 4), put numbers in the correct order (goal 3), and complete simple number-recognition activities (goal 1).

The movements accompanying the song are one way to add a kinesthetic element to learning, but the world is full of other fun ways for children to practice forming numbers. They can—

- write each number in the air while singing the corresponding lyric
- write each number on paper as they sing about it
- finger-paint numbers on glossy paper
- trace numbers cut out of sandpaper
- form numbers out of modeling clay
- draw numbers in a sand tray
- form numbers from pipe cleaners
- turn each number into an object they associate with it (as I did in the stylized numbers chart)

For a really different experience, half-fill a large self-sealing plastic bag with tempera paint. If the bag is placed on a solid surface, such as a table or the floor, children can write each number on the bag with a finger. The number will appear momentarily as the finger displaces the paint.

Other Kinesthetic Activities

The following ideas address number writing, recognition, and counting goals:

- Use wide masking tape to create very large numbers on the floor. Let the children walk on the tape, starting at the top just as they would if they were writing each number.
- Children can use their bodies to form numbers by lying down on top of the masking-tape numbers. They probably will have to pair up to make a whole number!

Number Song

Words and Music: Sarah Morgan

- Alternate between doing full-body skywriting (see principle 12 in chapter 2, pages 27–29) and writing the numbers on paper or any of the other materials listed previously.

Kinesthetic Strategies for Preventing Backwards Numbers

I've always told my students that the numbers 2, 3, 5, and 7 are "cave numbers." These numbers have to be written so that Big Bear—which is the index finger on their left hand—can crawl in when it is time to hibernate. Thus, they have to draw the shape of the number around their index finger as it lies on their paper. This strategy works only for students who are right-handed. For lefties, explain that their pencil point is Big Bear. When they are holding their pencil ready to write the number, they have to make the cave opening face toward the pencil point.

These numbers—2, 3, 5, and 7—are the ones most often reversed. This story provides your students with a mnemonic strategy and a tool for immediate self-correction.

Strategies to prevent number reversals for right-handers.

Strategies to prevent number reversals for left-handers.

For 4, 6, 8, and 9, demonstrate how they fit within the hand resting on the paper (see illustration). All four numbers begin with a movement toward the hand, followed by one away from it. Left-handed children can extend the thumb of the right hand as it holds the paper, then draw around the thumb to make these numbers (see illustration).

Number Recognition Games

Bingo: Give each child a small chart of the numbers to ten (see worksheet 4.3 on page 94) and ten poker chips. As you call out numbers in a random order, the children place a chip over each number you call.

Go fish: Divide the children into groups of about four. Give each group a deck of number cards (or a standard deck of playing cards with the face cards removed will work). Deal all the cards and explain that the goal is to make pairs of numbers. Each player in turn has to request a specific number from one of the other players.

Go grab: Make two sets of the numbers from one to ten. (You could write each number on an index card or photocopy the stylized numbers on pages 89–91 and cut apart.) Select 20 children and safety-pin a number to each child's shirt. Have the children line up in two teams, in order from one to ten, and stand facing each other about four feet apart. Toss a beanbag on the floor between the two groups while calling one number. The two children, one from each team, who are wearing that number run to the center and try to retrieve the beanbag. (With 30 children, make an extra set of numbers and have the children stand in a triangle or circle.)

Line up: Pin a number on each child's shirt as described in the previous activity. Divide the children into teams of ten composed of one child wearing each number. Give the teams a specific amount of time to arrange themselves into the correct sequential order. (The children who are not wearing numbers can act as timers and checkers for the teams.)

Stand up: Pin a number to each child. Sing the Number Song (page 43) in a group. As each number is named in the lyrics, have all the children wearing that number stand up. (It is not necessary to have an equal number of children wearing each number.)

Hopscotch: Outline a hopscotch grid on the floor with masking tape. (Make sure each square is large enough that children can actually step in it.) To play the game, a child tosses beanbags into squares with certain numbers in them, then hops through the squares in order from one to ten, skipping over any box with a beanbag in it. Have the class count as the child lands in each square.

Beanbag toss: Use the hopscotch grid described in the preceding activity. Designate one child as the caller who will shout out numbers. As each number is called, the players try to toss their beanbags into that square. (Play with a small group or set up a rotation system so everyone has a chance to play.)

Swift says: Use the hopscotch grid to play a Simon Says–like game with a small group. Give one child at a time a direction to follow using the number squares. Here are examples:

- Hop once on three.
- Sit on nine.
- Put your hand on five.
- Jump twice on six.

- Yawn on eight.
- Put both feet on four.
- Stomp on seven.

Assessments

This section suggests informal ways to assess achievement of the goals for this chapter. See Appendix B for reproducible tracking charts and progress reports. There are six reproducible worksheets in Appendix A that correspond to the assessments described in this section. When each child has completed all six pages, you could compile the work into a small booklet for the child to take home. (You may want to keep a copy of this booklet in each child's portfolio.)

Recognizing Numbers

Give each child a fives chart to ten with numbers supplied (worksheet 4.3, page 94). Call out directions to mark or color individual numbers with specific colors. Give directions slowly enough that the children have enough time to follow you but not enough time to become carried away with coloring. Sample directions follow, or you can make up your own, as long as each number is marked in a unique way, each direction involves only one number, and you do not call out the numbers in sequence.

- Put a circle around the three.
- Color the nine yellow.
- Put an X on the five.
- Draw a line under the seven.
- Make a box around the eight.

- Color the four red.
- Color the two blue.
- Color the one green.
- Color the six orange.
- Color the ten pink.

Counting to Ten

Listen to each child count from one to ten. If the counting is correct, check his or her name off your assessment list (Appendix B, page 174).

Ordering Numbers Sequentially

Give each child a page containing the numbers from one to ten in random order and a blank grid. Have the children put the numbers in correct order (worksheet 4.4a, page 95). Use worksheet 4.4b (page 96) for those children who are ready to order numbers to 20. They can either cut apart the numbers on the worksheet and glue them in the correct order on the grid, or simply write each number in its appropriate square.

Supplying Missing Numbers

Give each child a copy of worksheets 4.5a and 4.5b (pages 97 and 98). Have the child fill in the missing numbers to complete the number grid. There are five different choices, three pages for numbers to ten and two pages for numbers to 20, so you can use some pages for practice or repeat assessment.

Discovering Patterns

Give each child a five-frame chart to 20 (worksheet 4.6a, the top half of page 99). Have the children color as many patterns as they can find (for example, 1, 6, 11, 16).

Writing Numbers

Have children write the numbers to ten in their best handwriting. Allow them to choose a favorite color of crayon with which to write. (Use worksheet 4.7, page 100, or for children who can write the numbers to 20, use worksheet 4.6b, the bottom half of page 99.)

Extension Projects

You may wish to have children create personalized number books. These books could be as simple as pages of the stylized numbers, or there could be several pages for each number. For example, for the number three, the book might contain a page of the stylized numbers, a page of threes written by the child, a "found objects" page (such as a leaf with three lobes), a page containing a drawing of the child's three-member family, and a page of three fingerprints which the child has transformed into critters by drawing with a felt-tip pen.

The class could cooperate in making a set of ten books, one for each number. Each child could contribute one page and choose his or her favorite way of representing the particular number. Here are examples (refer to Gardner 1993 for explanations of the intelligences):

- A verbal/linguistic child could dictate a story about three of something.
- A logical/mathematical type may want to construct objects made from three shapes.
- A child strong in the naturalist intelligence may choose to find or draw objects from nature that have qualities of the target number (an ant with three body parts, for example).
- An interpersonal child may want to draw the members of his or her family (if they represent the target number) or draw the target number of close friends.

- A musical/rhythmic child could create a short poem, rhyme, or song about the number.

- A kinesthetic child might want to create a dance to go along with the song or rhyme the musical/rhythmic child wrote.

- An intrapersonal child might want to dictate a statement about what each number reminds him or her of, as a verbal "portrait."

- A visual/spatial child might be interested in drawing or painting original stylized numbers (different from those included in this chapter).

The ideas are endless, and the children themselves will come up with some of the best ones! It is wise to allow the children the choice of working alone or with a partner. The activities will result in many beautiful creations that will be admired and enjoyed for weeks. Parents certainly would enjoy seeing them displayed during an open house or parent/teacher conferences.

Chapter 5
Understanding the Meaning of Numbers

Goals for This Chapter

1. To visually imprint number meanings
2. To make visual connections for numbers to ten
3. To visually imprint computations to ten
4. To form visual imprints of where numbers are in relationship to one another
5. To recognize numbers within their context
6. To discover relationships between numbers

The magic ingredient in visual computing is the brain's ability to recall visual images. If you gain nothing else from chapters 1 to 3, the concept of visual imprinting is a must. In order for this method of computing to work for you, visual imprinting must take place—and it will, if you allow time to do some things that might appear at first glance to be a waste of time. The more care you take at the stage of the process described in this chapter, the more success your students will reap in future computation.

This chapter contains various activities to promote visual imprinting, visual connections, and pattern discovery. The results of this process will not be measurable using the objective assessments we have become accustomed to using. I will, however, detail how to tell when your goals for this stage have been reached. The process itself is so much fun that it may be tempting to rush through the activities, but please do not give in to this temptation! If you allow sufficient time to complete the visual imprinting exercises in this chapter, you and your students will see the payoff once you arrive at actual computation (described in chapter 7).

Visual Imprinting for Numbers to Ten

The objective for this section is for every child to emerge with a clear visual image for each number. You will not, of course, be able to dictate what that image will be—all that matters is that each child emerges with an image that is unique to his or her memory. The visual image will probably involve a dot pattern and may include some color. You will have a chance to see some of these images as the children produce them for you in their final projects.

Initiating the Number Imprinting Process

1. Choose a target number. Set up the overhead projector and place that number of poker chips on it so the whole class can see them.

2. Ask the children "How many are there?" Do not allow them to count; instead, ask them to guess! (We want this to be an instinctive, visual imprinting process, not a cerebral one.)

3. Quickly rearrange the chips and ask again, "How many are there now?" Repeat this step several times. You want the children to understand that no matter what the arrangement, the total number stays the same.

Materials Needed
- overhead projector
- poker chips or similar tokens

4. Ask questions to elicit that changing the location or arrangement of the chips does not change the quantity. For example, "If I put these on the floor, how many are there then?" or "If they are on the ceiling, how many are there?" "If I hold them tightly in my hand, how many are there?" Students will remember the point.

5. Place one dot grouping for the target number on the overhead projector. Ask, "What does this look like to you?" (Visual connections are being made.) Let volunteers give answers. Then show another grouping and ask the question again. Feel free to tell students what that dot pattern reminds you of. Here are some ideas to get your creative juices flowing:

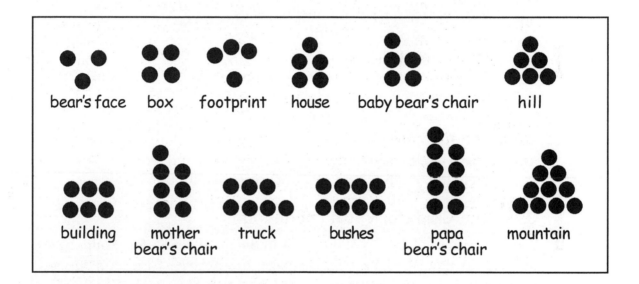

bear's face   box   footprint   house   baby bear's chair   hill

building   mother bear's chair   truck   bushes   papa bear's chair   mountain

Imprinting for Odd Numbers

When you are teaching odd numbers, show a poker-chip arrangement in which all the chips but one are paired and say, "This reminds me of children walking with partners to the library" (or some other similar image). Ask, "Does every child have a partner?" When the students respond no, give each student or small group of students the target number of chips. Let students attempt to arrange the chips in a different way so that each "child" has a partner. If they decide that they are unable to make every child have a partner, suggest that the extra child must feel "odd." Discuss what the children might do to keep that child from feeling left out. We hope one of them will suggest taking turns! (This is a real-world connection that enhances memory of the concept.)

Making Dot Cards

1. Using paper marked with a grid pattern, model how to make the target number of dots in each square by dipping that number of fingertips in the paint, then pressing those fingers in the square. Use a different arrangement each time. (You need not always use the same fingers, just the same number of fingers.)

2. Instruct students that each square in the grid should have a different grouping for that target number (visual imprinting). Stress creativity of design rather than speed!

3. Let the children talk about what they did and what they were thinking when they made the various patterns (visual connections).

4. When all the papers are dry, laminate them and let the children cut them into cards for use in the activities that follow.

Materials Needed

- tempera or watercolor paints
- 1 sheet of sturdy paper per child, with a grid pattern on it *(For fastest preparation, photocopy worksheet 5.1, page 101, on both sides of each sheet.)*
- fingers
- old clothes or painting shirts
- paper towels
- scissors
- lamination equipment or clear plastic shelf paper

Using Dot Cards for Visual Imprinting

You may wish to place sets of dot cards in the math center for groups to use during math time. Worksheets 5.2a–b (pages 102–3) have a basic set of dot cards or, preferably, let students share the dot cards they have made with the class.

Dot flash cards: Divide the students into pairs. One child displays a card and the other says what number it represents. (Children take turns showing and naming the cards.) Using the student-made dot cards is best, or use worksheets 5.2a–b.

Bingo: Photocopy the dot card bingo boards. (There are two different boards, labeled worksheets 5.3a and 5.3b, pages 104–5.) Give each child a board and a set of chips. One child calls out the numbers while the others place chips on their boards.

Memory: Using two sets of dot cards, shuffle the cards and arrange them face down. (Use the children's dot cards or worksheets 5.2a–b.) A pair or small group of students can play memory. Each player turns over two cards in an attempt to find the two with the same number. (Have the player name each card.) If the cards have different numbers, the child must turn over both cards and the next player takes a turn. If the cards match, the player keeps them and takes another turn.

Go fish: Using four sets of dot cards, students can play go fish in groups of up to four players. Players must name the card they wish (rather than pointing or showing a card). If children use their own dot cards, then they will just be matching the cards with different arrangements of the same number of dots, rather than identical cards.

Floor dominoes: Reproduce, laminate, and cut apart the floor domino cards (worksheet 5.4a–f, pages 106–11). Let the children play in pairs or small groups, naming the numbers as they play.

War: Children play in pairs. Each player shuffles his or her dot cards and holds or places them face down. On a signal (such as, "1–2–3–War!") both players put a card down in the center at the same time. Whichever child has the larger number takes both cards. Alternatively, players can name a number before turning their cards over. If that number shows up, the player who said that number takes both cards. If not, the center pile builds. The game may also be played with naming "odd" or "even."

Visual Imprinting for Computation

The objective of this section is for the children to develop such a rich visual background for computation that they will "see" answers in their heads when confronted with abstract problems on a page. At this point, however, the process is quite passive. You will just be giving the students opportunities to look and discover, without actively learning math facts. The process of building visual images in children's memories is analogous to taking pictures of various groupings of dots. Until you develop the film, you cannot see the images, you just have to trust they are registered on the film.

Imprinting Computations

1. Copy the two pages of dot cards on worksheet 5.2a–b onto transparency film and cut apart.
2. Place one card on the projector and ask "How many are there?" Repeat this procedure with each of the cards.
3. Give each child a marker and paper. Explain that the students will be

Materials Needed

- transparency film
- overhead projector
- worksheets 5.2a–b (pages 102–3)
- scissors
- marker and paper for each child

reproducing on their papers what they see on the overhead. Use the following process for each number:

- Show the card.
- Remove the card.
- Have students close their eyes and "see" the card in their minds.
- Have students open their eyes and draw what they saw.
- Show the card again and have everyone name the number in unison.

4. Show a card on the overhead projector. Model a description of how the dots are grouped on the card and have the students repeat in unison. For example, "Two and one more is three" (see illustration).

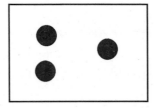

Using Dot Cards for Computation Imprinting

At this level, it is best to use identical pairs of cards, so I recommend making two copies of worksheets 5.2a–b, laminating them, and cutting the pages apart into cards.

Memory: This time when the children play memory, they will identify what they have paired by identifying the grouping, such as "one and two more is three." The objective here is to have the child move beyond counting to recognition of quantities as a prelude to addition. For the larger numbers, there is often more than one reasonable visual grouping. For example, the set of five dots in the bottom left card of worksheet 5.2a could be four dots in a square plus one more, or three dots in the top row plus two in the bottom. Any grouping that totals the target number of dots is acceptable.

Go fish: Players ask for a dot grouping they want, such as, "Do you have a one-and-two-more-is-three?" Any card with three dots qualifies.

Visual Imprinting for Number Relationships

The activities in this section will lay the foundation for children's understanding of where numbers fall in relationship to each other within the global sequence of numbers. Just where is three in relationship to five? How much bigger is three than one? And how many more do you need to get from three to five?

Materials Needed

- pyramid dot cards (worksheets 5.5a–b, pages 112–13)
- five-frame dot cards (worksheet 5.6a–b, pages 114–15)
- photocopier
- lamination equipment or clear plastic shelf paper
- scissors

Make one photocopy of both worksheets for every two children. Laminate the worksheets and cut apart.

Pyramid Dot Cards

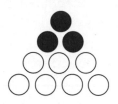

The patterns in the pyramid dot cards show how each number relates to ten. In this example, we can see a positive (black) three, a negative (outlined) seven, and ten as a whole. We also see three being made from one and two, and seven from three and four.

Five-Frame Dot Cards

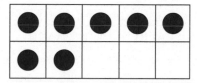

This pattern anchors numbers to five and ten. This is seven. It is also five and two more, or "a hand and two fingers." It is also three less than ten.

Using the Pyramid Cards

Memory: Play in the traditional way except that children must describe each card as they turn it over, in terms of how many black dots, how many outlined dots, and how many total dots are on the card. For example, a pyramid with three black dots and seven outlined dots could be described as "three and seven more makes ten" *or* "seven and three more makes ten."

War: The children must label each card they lay down in terms of how many black dots, how many outlined dots, and how many total dots are on the card. For example, a pyramid with three black and seven outlined dots could be described as "three and seven more makes ten" *or* "seven and three more makes ten," depending on whether children want to play that the number of black dots or the number of outlined dots determines who wins.

How many: The first child turns over a card and asks the second, "How many (black) dots are there?" The second child answers and asks, "How many to ten?" The first child answers (by identifying the number of white dots). Then the second child turns over a card and the game continues.

Using the Five-Frame Cards

1. Begin with number-recognition activities. Show a card and ask, "How many are there?" Do this activity as a class first, then let the children pair off and play together.

2. Next, make a connection to children's hands. First ask how many fingers students have on one hand. Then ask how many boxes there are in the top row of the fives frame. Make a solid connection between the fingers of the left hand and the boxes in the top row of the chart. Next, make a solid connection between the right hand and the second row.

3. Hold up a card or place a transparency of the fives frame on an overhead projector. Ask the children how many dots there are, then ask them to show you on their hands how many.

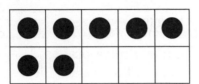

This frame depicts seven. To show it on their hands, the children would hold up all five fingers of the left hand and two fingers of the right hand while saying in unison, "Seven."

4. Use the cards to identify the relationship between a number and its anchor numbers. (Anchor numbers are multiples of five, the numbers at the beginning and end of the frame.) Do this as a class until the children are fluent with the activity, then let them pair up and practice. For example, given the preceding five frame (which shows seven), ask questions such as these:

    *   How many are there?
    *   How many more than five? (This is an example of using an anchor number.)
    *   How many bigger than one is this?
    *   How many more to ten?

    To go one step further ask, "What if I put two more dots on? How many then?" You will hope to see that at least some of the children can figure this out with-out counting.

Pattern Discovery

The purpose of using a large five-frame chart is that it places the numbers in context, so that the relationships between them are easier to see and it is easier to discover patterns within the global whole. Using the five-frame chart (see illustration) will also provide a foundation for multiplication and division later on.

1	2	3	4	5
6	7	8	9	10
11	12	13	14	15
16	17	18	19	20
21	22	23	24	25
26	27	28	29	30
31	32	33	34	35
36	37	38	39	40
41	42	43	44	45
46	47	48	49	50
51	52	53	54	55
56	57	58	59	60
61	62	63	64	65
66	67	68	69	70

1. Make one copy of the five-frame chart to 70 (worksheet 5.7, page 116) for each child. (Note that page 116 has two charts on it.)

2. To begin, allow the children plenty of time to explore the number chart. Pattern discovery is enhanced if children are allowed to talk to each other about what they see.

3. Students could use colored pencils or highlighting markers to color in the patterns they discover. As discussed in chapter 4, each column forms a repeating pattern (see page 41). There are other discoveries to be made as well.

4. Let small groups or the entire class collaborate to make a really long chart (use the blank charts on worksheet 5.8, page 117), the scope of which they will determine. They may wish to find out if the patterns they discovered will keep on going or will stop at some point.

5. Lead a "what if" game, using the five-frame chart and poker chips, to help students predict the patterns. For example, you might say, "Put a chip on five. Now count five more and put a chip on that number. What if you kept counting five more and putting a chip on each of those numbers? What do you think you would see?" Encourage student predictions. Now see what patterns develop when you add other numbers repeatedly. Examples for four, eight, and 12 are shown on page 58. Because the objective is strictly to identify visual patterns in the number grid, not to memorize specific sums, you can start at any number in the grid.

6. Experiment to see what patterns develop when you don't add the same number each time, but rather add a repeating pattern of numbers, such as four, then five, then six, then four… (This pattern is shown in the second five-frame chart on page 58.)

7. Children who are very visual may continue finding patterns as they use this chart for discovery. For example, they might note that if they add five to any number, they get the number right under the number they started with. If they take five away from a number, they will end up with the number right above the number they started with. If they add ten, they will jump down exactly two rows. Also note that adding moves one down the chart, whereas subtracting moves one upward. Some highly visual children will apply patterns such as these in solving arithmetic problems, as exemplified in the case study of Ethan on page 23. The examples that follow show various patterns:

1	2	3	4	5
6	7	8	9	10
11	12	13	14	15
16	17	18	19	20
21	22	23	24	25
26	27	28	29	30
31	32	33	34	35
36	37	38	39	40
41	42	43	44	45
46	47	48	49	50
51	52	53	54	55
56	57	58	59	60
61	62	63	64	65
66	67	68	69	70
71	72	73	74	75

The first four rows of this chart show the stair-step pattern that results from adding (or subtracting) four repeatedly. I chose to start with five because that gives the clearest illustration of the pattern. The rows starting with 32 show adding (or subtracting) six. Some children may realize that the "shape" of adding four will be a slant to the left, while the "shape" of adding six is a rightward slant.

1	2	3	4	5
6	7	8	9	10
11	12	13	14	15
16	17	18	19	20
21	22	23	24	25
26	27	28	29	30
31	32	33	34	35
36	37	38	39	40
41	42	43	44	45
46	47	48	49	50
51	52	53	54	55
56	57	58	59	60
61	62	63	64	65
66	67	68	69	70
71	72	73	74	75

This pattern is the sequence of starting with three, then adding four, then five, then six, repeatedly (described in step 6 on page 57).

1	2	3	4	5
6	7	8	9	10
11	12	13	14	15
16	17	18	19	20
21	22	23	24	25
26	27	28	29	30
31	32	33	34	35
36	37	38	39	40
41	42	43	44	45
46	47	48	49	50
51	52	53	54	55
56	57	58	59	60
61	62	63	64	65
66	67	68	69	70
71	72	73	74	75

This chart shows the pattern formed by adding eight repeatedly. One picture that emerges is that adding eight involves moving down two rows then left two numbers (that is, add ten, subtract two), an example of using ten as an anchor number. (Using five as the anchor number, this pattern could also be described as down one row and three to the right; that is, add five and three more.)

1	2	3	4	5
6	7	8	9	10
11	12	13	14	15
16	17	18	19	20
21	22	23	24	25
26	27	28	29	30
31	32	33	34	35
36	37	38	39	40
41	42	43	44	45
46	47	48	49	50
51	52	53	54	55
56	57	58	59	60
61	62	63	64	65
66	67	68	69	70
71	72	73	74	75

Adding 12 is similar to adding eight. It involves moving down two rows and right two numbers. This pattern occurs because 12 is ten and two more (or one ten and two ones).

8. Experiment with adding various numbers, not by counting up, but by establishing and continuing the pattern. For example, as illustrated, eight is "a row and three more" (and later will be "a hand and three more"). To add eight to a number, add a row by simply going straight down, then count to the right three places. Other numbers with which to experiment include adding by 15s, 10s, or 20s.

This time of discovery is but *one pathway out of several* that will enable your students to make sense of addition and subtraction. It takes several approaches to complete the picture for the whole class. Some children respond quickly to the five-frame charts, some to "my two hands" (chapter 6), and others to the patterns they discover when they learn the number families (chapter 7). If some students do not catch on to the five-frame chart, seek a different strategy that makes sense to them.

The point is that discovery time is just that: time for children to try the various activities and to discover for themselves which methods they feel the most comfortable with. The value of doing all the activities with the whole class is that even though each child may emerge with one favorite method, having experienced several visual-kinesthetic approaches will only add richness to the student's experience with numbers.

Becoming Familiar with Numbers to 20

To familiarize the children with the numbers to 20, use just four rows of the five-frame chart. (See worksheet 4.6, page 99, for a five-frame chart to 20 and a blank frame.)

1. Let students write the numbers to 20 into a blank frame.

2. Let them talk about the patterns they remember, if they wish to do so.

3. When the children seem familiar with the number frame, provide each with another blank frame and a set of poker chips. Play a bingo-like game in which you call out a number and the children place a chip on the blank space where that number would be found. Alternatively, as you call out each number, you could have students write it in the space where it belongs. Another variation is to identify a number by location rather than name. For example, "Put a chip on the number under three." "Put a chip on the number over nine." Your goal in these activities is to help the children become very familiar with each number's location on the "global map," its "address" if you will. When the fun of playing these games has begun to fade, it is time to move on into computation. I'm ready, aren't you?

Final Projects

"When You Say . . . I See . . . " book: To make the booklet, cut sheets of paper in half, then fold five half-sheets in half. Place a cover page (see page 118) on top of the half-sheets and staple them all together on the fold to make a booklet with ten inside pages. Each student can write his or her name on the front cover, and decorate if desired. On each inside page, they write one number in order from one to ten. Have all the children turn to page 1, and ask, "When I say 'one,' what do you see?" Let each child draw his or her visual image of the number. Repeat for each number up to ten. (I would not recommend doing all the numbers at one sitting unless the students really want to.)

"Dot Critters" book: This book starts out looking like a "When You Say…I See…" book. However, it uses dot patterns to represent each number. When the children have written the numbers and dot patterns, they take fine-tip markers and add legs, antennae, spots, eyes, smiles, or whatever to turn the images into critters. To add interdisciplinary learning, each child could dictate a story about the critters. For example, perhaps there is one more critter on each page because they are going to the playground and gather friends along the way. (A reproducible cover for this book is provided on page 119.)

Assessments

Conduct informal assessment by observing the children daily during their partnered activities. Make notes to yourself about what you observe, including particular children's preferred method of remembering. Which children prefer the five frame and which the pyramid dot cards? Do they recognize dot patterns with ease, or do they have to think for a second before answering? Progress charts relevant to the skills in this chapter are provided in Appendix B.

The following procedure should not be used for giving grades, but rather serves as an assessment of students' learning thus far. You will be able to identify any weak spots in teaching or areas where the class needs practice. Make the assessment as low-key as possible; for example, don't describe it to the children as a test. The procedure is as follows:

- Choose a set of five-frame dot cards on overhead transparencies.
- Have the children number their paper according to how many cards you have.
- Place one card on the overhead at a time. Ask the children to look, close their eyes and visualize the numeral that corresponds to the number of dots on the card, then write the numeral on their paper.

You may repeat this assessment as needed to check for progress.

Chapter 6
Making the Transition from Visual to Symbolic

Goals for This Chapter

1. To make the transition from visual to symbolic representation
2. To understand the action of computation
3. To provide a mind/body tool for computation

This chapter continues the visual-kinesthetic approach we have begun but makes a natural transition from visual to symbolic representation. The way I like to do this is through a storytelling approach. Stories allow the child to engage in the process, show the real-world applications of math, and begin to create an understanding of what is happening during computation. By this I do not mean memorizing a formula, but seeing and doing the action of computation.

Story Problems

Twitch: My favorite series of stories involves a precocious squirrel named Twitch. I have spent some lovely hours relating to my class the misadventures of this character, who is nine parts rash to one part cautious. Twitch's goal in life is to collect acorns, which he gathers in a little bag slung over his shoulder. Twitch lives in a hole up a tree in a neighborhood much like your neighborhood. There is a huge calico cat named Irving living on one side, and a hound named Chunk on the other. These neighbors add some twists and complications to the hunt for acorns. There have been times when Irving has pounced upon Twitch and caused him, in his fright, to drop acorns. At other times, Twitch, believing Chunk to be sound asleep, creeps too close to grab an acorn. Chunk wakes up, and the chase is on. As I tell a story about Twitch, the children pick up the number of acorns that Twitch finds, and put back any he drops. I purposefully use very small numbers so as not to divert a lot of attention to the counting. (The children might have little cloth bags and real acorns that have been coated with shellac or polyurethane to protect against rot and bugs—or at least small pebbles to represent acorns.) The children enjoy recapping the story when I have finished. They dump their acorns out on the table in front of them and talk about how many Twitch might have had if he hadn't dropped any!

Pool party: Children at a pool party have a variety of reasons for getting into and out of the pool. The children can draw pools on paper and then use miniature plastic bears or other markers to represent the children in the story. At first, they can just practice taking the right number of children out of the pool and putting others in. As they do so, they will be feeling in their bodies the movements of adding and subtracting as motions of "bringing in" and "taking away."

At the playground: Another possible setting is a playground at school (or another setting where children like to be or go). The focal point is a long swing set with children coming and going for various reasons. Again, have the students draw a representation of the swing set and give them materials to represent the children in the story.

Using Story Problems in the Transition to Symbols

Once the students are really clear on the movements behind adding and subtracting, the next step is to have them write equations as you tell the story. Begin by explaining the symbols in an equation:

- The first number in the equation is the number you have at the beginning of the story.

- The symbol (+ or −) shows that something happened: either some went away (make a horizontal motion with your arm that resembles the shape of the minus sign) or some came to join the first number (cross your arms over your chest to resemble a plus sign).
- The next number tells us how many either came (+) or went away (−).
- The = sign is the signal that we are at the end of the story and need to stop and take a picture of how many are left now.
- And the last number tells us what was left.

At first, talk about and model how to write the problem as you tell each story. Eventually, the children will join you in writing the problems during the story. As an example, if three children are in the pool, and two more get in, the expression is $3 + 2 = 5$. As you write this, you would use language such as this, "Three children were in the pool (point to the three) and here come (cross your arms over your chest) two more! Stop for a picture (point to =). There are five all together at the end of the story."

If you tell them next that one child wanted a drink and got out of the pool, the students would write $5 - 1 = 4$. You would explain: "There were five children in the pool. Oops! There goes (slicing arm motion) one child! Stop and take a picture. There are four left in the pool at the end of the story."

Story Problem Activities

For a gross motor approach to story problems, the children can act out the equation instead of writing or moving small objects.

- If the story is about a pool party, use masking tape to mark off the boundaries of the pool and have children actually jump into or climb out of the "pool" while you tell the story.
- If the story is about the playground, arrange chairs in a line to represent a swing set, and have children actually come and go as you tell the story.
- Another idea is to place a tape recorder and an audio tape of short recorded stories in the math center. Then children can practice computation on their own or in pairs during their time in the math center.

"My Two Hands"

At long last, here is the magical strategy that started Alice humming with computation. The skill it builds is that of enabling the children to see the sums, rather than counting up to them or memorizing answers. The reason to avoid counting up in computation is simply that if children learn this way, they will never progress beyond counting on fingers or on touch points on the numbers—a practice that is very common in our schools. To compute visually is to compute quickly.

To introduce "my two hands," hold up two fingers and ask, "How many?" Spread the same fingers wide apart and ask again. Then, with your fingers spread, say slowly and deliberately, "One plus one is two" (see illustrations).

Two and no more.   One and one is two.

Repeat this process for three. Say, "Show me three fingers!" Students will hold up three fingers. Then say, "Show me three another way." Guide them, if necessary, into showing a visual of 1 + 2. Describe this as, "One and two more is three."

Three and no more.   One and two is three.

Four is the four fingers on one hand. Repeat the same process as before, using the same language. Have the children watch and imitate you as you move your middle finger from one side to the other to make 2 + 2 and 1 + 3 (see illustrations).

Four and no more.   One and three is four.   Two and two is four.

Five is an important number. Five is all the fingers on one hand, and five is also one row on the five-frame chart. Five is the number you add to for numbers over five, and the number you take from for numbers under five. It is a great anchor number. The sum 1 + 4 is represented with the thumb and four fingers. Two short digits and three tall ones make 2 + 3.

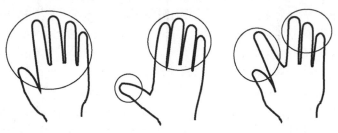

Five and no more.   One and four is five.   Two and three is five.

Six is the fingers of one hand and the thumb of the other. The sum 1 + 5, is described as a hand and one more. For 2 + 4, you have two thumbs and four fingers. For 3 + 3, you have three short digits and three tall ones.

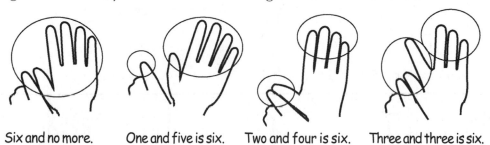

Six and no more.   One and five is six.   Two and four is six.   Three and three is six.

Continue in this fashion for all the numbers to ten. Use your hands to show the various ways to make each target number. If the child begins to count up to determine the number of fingers you are holding up, you have a teaching moment on your hands. Ask him or her to guess what the number is.

Extension Activity

Once the children can "see and do" the sums to a particular target number, show them why particular combinations are the same. Take 1 + 2 and 2 + 1, for example. Hold up your three fingers to show 1 + 2 like this: ▶

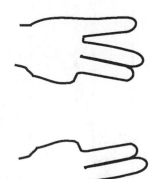

Then, while the children watch carefully, flip your hand, without moving your fingers at all, so that now the fingers show 2 + 1 like this: ▶

Talk about how the fingers stay just the same, but they trade places. I like to do this part of the lesson holding my fingers horizontally since this visual will closely reflect the "families" introduced later (see chapter 7).

Tough Spots

Sums for seven and eight seem to require the most practice. It helps quite a bit to use the children's own hands to help them see 3 + 4, 2 + 5, 2 + 6, and 3 + 5. I would use the five-frame dot cards as additional visuals with the sums that cause difficulty. Remember to use each difficult sum as a teaching moment and an opportunity for additional practice. You can demonstrate to the class the relationship between the "my two hands" method and the five-frame chart by holding out your hands, palms nearly flat, one over the other in order to show a row and several more. They would look something like this:

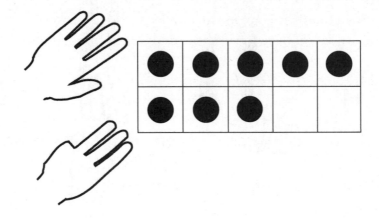

Making the Transition from Hands to Mind

Use your hands: When I first introduced this concept to Alice, she quickly understood what was happening. I was delighted to watch her looking at her hands as she was doing her practice sheets. After a while, however, I feared that this practice could also become a crutch, although not as limiting as counting on one's fingers.

See it in your mind: So I took Alice to the next level, which was to ask her if she could see her hands doing the movements in her head. She was game to try. She would squeeze her eyes shut and, after a moment, she would exclaim, "I can see it!" She must have, too, because she would quickly open her eyes and write the answer. It was not long before Alice rarely had to stop to picture a computation in her head.

Chapter 7
Mastering Computation to Ten

Goals for This Chapter

1. To create a real-world context for computation
2. To provide a global view of all number families
3. To conduct pattern discovery within number families
4. To master computation to ten

Now that students have a rich and vivid background laid for computation, let's get to it! Welcome to Stony Brook Village, the strategy I use to help students visualize computation! Let's take a walking tour before we start to work. Coming up is a global view of the residential part of town.

Orientation to Stony Brook Village
Property Managers

You will be assigning students the duty of being property managers, so you will need to explain to them what this means. "*Property,*" I tell them, "is all the stuff you can see: the houses and the land. Being a *manager* of all this property means that you take care of it." I explain that the job of the property manager is to keep track of who lives in the houses on each particular street. Each street in town has a separate property manager, and if a family moves out, the manager has to rent the house to someone else. The most important job of the property manager is to oversee how many people can live on each floor of the house. Even five-year-olds understand and use this title with ease. In addition, the title of manager reflects the children's own management of their learning process.

Streets and Houses

As you can see from the map on page 69, each street in the residential section is clearly marked with a sign. The name of the street matches the number written in the attic of each house on that street. The rule in this burg is that only that many people can live in the house. So, on Third Street, only three people can live in each house. On Fourth Street, only four people can live in each house, and so on. We don't know for sure how these rules were established, but we do know that over time, the rules got more and more involved and specific.

Apartments in the Houses

The next rule the Planning Commission made was that there were to be two separate apartments inside each house, so that each house could shelter two families, one on each floor. (The Planning Commission is a group of people who have meetings and sit around a table and make rules for the town. They like things done a certain way, and the property managers have to obey their rules!)

Apartment Residents

That rule was apparently not enough, because the next rule the commission made was that no two houses could be populated in the same way. (This means that no house could have the same pattern of residents as another house.) For example, if you look at Fourth Street, you will notice that the first house on that street has no one living upstairs, but a family of four downstairs. The second house has to be different. It has one person living alone upstairs and three people living downstairs. In the last house, two live upstairs and two live downstairs.

The First House on a Street

We are not certain if a rule was made to this effect, but the fact is that property managers always seem to leave the top floor of the first house empty if they have a family that has the maximum allowable number of residents. For example, if a family of four wanted to live on Fourth Street, the property manager would put them in the first house, on the bottom floor, so that the empty apartment could be at the top of the first house (see illustration).

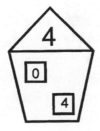  

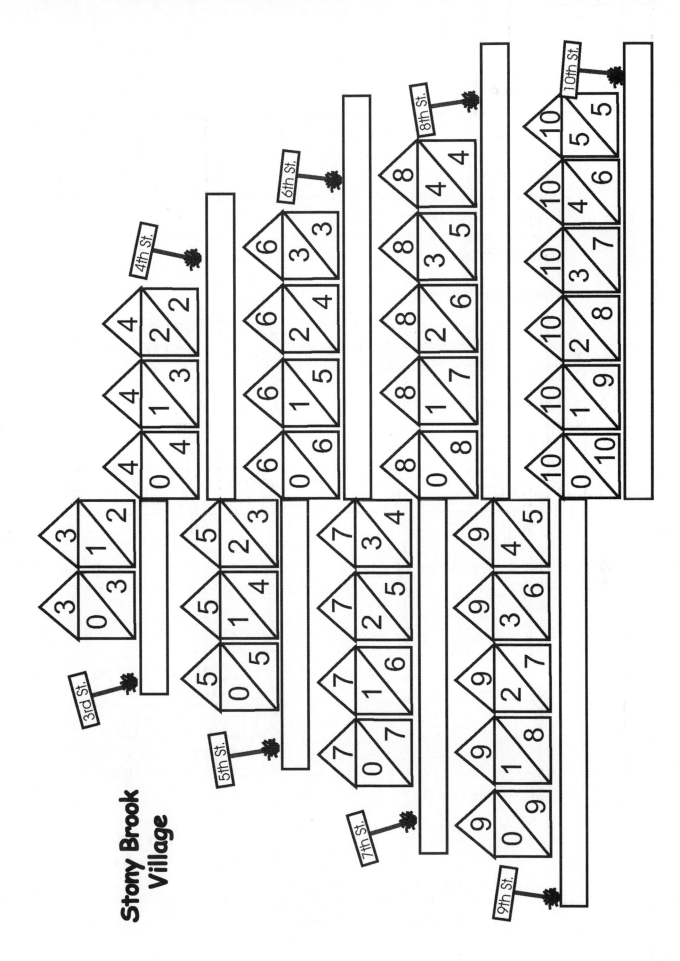

Stony Brook
Village

Addition and Subtraction, © 2002 Zephyr Press, Chicago • www.zephyrpress.com

Moving

The humorous side of all this rule making is that once the families have lived in their apartments for even a day, they are allowed to switch apartments with the other occupants of their house. They may not, however, switch houses altogether, as this would upset the delicate balance of the population. If one of the two families in the first house tried to trade places with a family in the second house, there would not be the right number of people living in either house anymore. The property manager would get into trouble with the Planning Commission, and everyone would be sad.

Pattern Discovery

Copy the village map (page 69) onto transparent film and project it on an overhead, or give a copy to each student. Guide students to notice interesting facts about the distribution of people in the homes.

Odd and Even Houses

- The even-numbered houses and streets are all on the right side of the map.
- The last house on each even-numbered street has an equal number of residents per floor. For example, the last home on Fourth Street has two residents living upstairs and two downstairs. If you look at the last houses on Sixth, Eighth, and Tenth Streets, their residents are evenly divided also.
- Odd-numbered streets are on the left side of the map.
- The house at the end of each odd street has one more resident on the bottom floor than on the top.

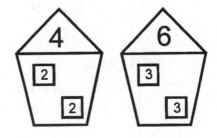

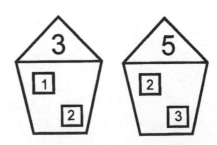

Comparing Streets

- Choose two streets, say Fourth and Fifth Streets, and do some comparisons. Ask, "How many houses are on each of these streets?" (Three houses on each street.) "Why do you think there are the same number of houses when five is bigger than four?" "What is different about how the numbers are placed in the houses on these two streets—particularly in the last house on each street?"

- Repeat this process with Sixth and Seventh Streets, noting that Sixth Street is an even street, so there are an equal number of people living on each floor of the last house (namely, three people on each floor).

- Seventh and Sixth Streets are remarkably similar. Take time to note that the top floors of the houses on those two streets are identical. The bottom floors of the houses on Seventh Street contain exactly one more person than the bottom floors of the houses on Sixth Street. Ask, "Why is this the case?" (The attic number on Seventh Street is one digit larger than the attic number on Sixth Street, and the extra person lives downstairs in each house.)

- Repeat these comparisons with Eighth and Ninth Streets.

Number Sequences

- Note the sequence of numbers in the last houses on Fourth, Sixth, Eighth, and Tenth Streets. The numbers of residents form this pattern: $2 + 2$, $3 + 3$, $4 + 4$, and $5 + 5$.

- Compare that pattern with the pattern of residents in the last houses on Third, Fifth, Seventh, and Ninth Streets: $1 + 2$, $2 + 3$, $3 + 4$, $4 + 5$.

Houses on Streets

- Note the number of houses on each street. For even-numbered streets, the number of houses equals half of the target number plus one. Thus, on Fourth Street, there are half of four (that is, two) plus one more to total three houses. Fourth and Fifth Streets have the same number of houses because half of five is the same as half of four, except for one extra person, who lives on the bottom floor.

- See if the children find any additional patterns or points of interest, and record their findings.

Summary of Procedure for Teaching Number Houses

Work through *all* the following steps for *one street at a time* before presenting the next street, starting with Third Street. Worksheets 7.4 to 7.10 refer to series of worksheets (distinguished with the letters a, b, c, and so on). In each cycle through these steps, present only the version of the worksheet that is appropriate for the number family you are targeting.

Discovery

1. **Felt houses and bears:** The children arrange small toys, such as plastic bears, in felt number houses, in order to discover with concrete objects the combinations that total the target number (worksheets 7.1 and 7.2a–b).

2. **Smallest to largest:** The children place number cards in the windows of the number houses, thereby discovering the pattern that, on each street, the smallest number goes with the largest one, the next smallest with the next largest, and so on (worksheets 7.1–7.3b).

Exploration

3. **Discovery worksheet:** Worksheet 7.4 mirrors the discovery process in step 2, but now the children write the appropriate numbers in the house windows.

4. **Practice houses worksheet:** Worksheet 7.5 has students repeatedly write the pairs of numbers for each street.

5. **All and no more:** Students examine worksheet 7.5 to determine that there are no other possible combinations that total the target number.

6. **My two hands:** Children make the number combinations for the street on their hands, to build a mind/body/visual connection to the number pairs for that street.

7. **Practice strips:** Worksheet 7.5 is presented again, this time with houses partially completed in a random order, rather than the strict numerical order of the Stony Brook street.

8. **Families went walking:** Worksheet 7.6 introduces subtraction from the target number.

9. **Additional practice:** If any students have not completely mastered the tasks presented so far, provide additional practice.

10. **Mixed addition and subtraction practice:** Worksheet 7.7 contains mixed addition and subtraction problems for the target street.

11. **Which house?** Worksheet 7.8 is an optional worksheet that makes the transition from number houses to traditional addition and subtraction problems.

12. **Writing problems:** Worksheet 7.9 is another optional worksheet that has children translate each house on the target street into the traditional addition and subtraction problems that correspond to it.

13. **Mixed numbers practice:** Once students have mastered at least Third and Fourth Streets, you can present worksheet 7.10, which mixes problems for the target street with problems for the previously mastered streets.

14. **Additional practice:** Worksheet 7.11 is a page of blank number houses you can use to make up problems for any student who needs some extra practice before moving on.

Discovery

After discovering patterns for the whole village (the global approach), it is time to start working at the beginning and proceed in steps. The following steps are for each attic number from three to ten. *Work through all the steps for one number at a time, starting with Third Street and ending with Tenth Street.* (Note that there is no Second Street because in my experience, most children already know 1 + 1 and 2 + 0.) My rule of thumb is that students generally can work through steps 1–12 for a single street in one practice session.

1. Felt Houses and Bears

Students begin by using concrete objects to discover the combinations that total the target number. I like to use small plastic bears in felt number houses. Distribute to the children the correct number of houses they will need for the street you are working with. For example, for Third Street, each child will need the following:

- two houses
- two 3 number cards for the attic
- bears or tokens to fill both houses, six in all

Have the children place an attic number in each attic, so they remember how many bears can live in each house. Review the rules for populating the houses: The total number of bears (or tokens) must equal the attic number, and no two houses can have the same combination of residents in its apartments. Let the children figure out how to arrange the bears to meet these criteria.

Directions for Making Felt Number Houses

Worksheets 7.1 to 7.3b in Appendix A (pages 120–24) contain patterns for felt houses, attic numbers, and window numbers. Follow these directions for preparation:

Worksheet 7.1. Felt-house pattern: To make a template, place the pattern on a piece of stiff cardboard. Cut around the outside of the house and cut out the door and windows. Place the cardboard template on a piece of felt and cut *around the perimeter of the house only.* Use a marker to draw the windows, door, and attic line on the felt.

Worksheet 7.2a. Attic numbers—page 1: Make *one* photocopy of this page, laminate it, and cut apart the attic numbers. Place a piece of self-adhesive Velcro® hook-and-loop fastener on the back of each attic number so it can be stuck to the felt house. (Rolls of Velcro® are available in fabric stores.)

Continued

Worksheet 7.2b. Attic numbers—page 2: Make *two* photocopies of this page, then follow the directions given for page 1 to make laminated attic pieces to stick on the felt.

Worksheet 7.3a–b. Window numbers—pages 1 and 2: Make *two* photocopies of each of these two pages. Laminate the pages then cut apart the numbers. Place a piece of self-adhesive Velcro® hook-and-loop fastener on the back of each window number so it can be stuck to the felt house. There will be enough windows to make one town, plus a few extras. (These are used for step 2, following.)

Quick preparation: If the preparation time to make felt houses for all your students is prohibitive, you have two options. You might simply photocopy the house pattern, attic numbers, and window numbers; laminate all the pages and cut apart the attic and window numbers; then place Velcro dots on both the house pattern and the numbers, so the numbers can be stuck to the houses. Another option is to make only enough felt houses for a small group to use and to do this step with small groups in the math center, rather than with an entire class simultaneously.

2. Smallest to Largest

Next, put the bears away but have children keep the felt houses and attic numbers. Give them laminated window numbers. Have them line up the window cards in a row, starting with zero and finishing with the highest number they have. Then instruct them to figure out which pair of numbers goes together in each house, reminding them to use all the numbers. (Don't forget that for even-numbered streets, students will need two copies of the middle number in the series; that is, two 2s for Fourth Street, two 3s for Sixth Street, two 4s for Eighth Street, and two 5s for Tenth Street.) When all the students have finished, talk about what they have discovered. As shown in the illustration, the solution is to pair the first and last number in the row, then the next two numbers in, and so on. By picking up and placing actual numbers,

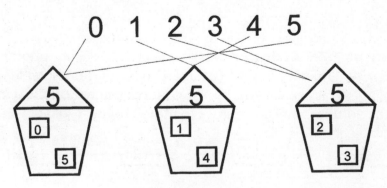

the children will feel the rhythm and movement of starting from the outside and moving in, particularly if they grab two numbers at once, one in each hand, to place in the windows.

Street Patterns

At this point, you will want to guide the children to arrange the houses in a meaningful pattern so they will be able to remember the whole picture without having to memorize the numbers. It would not be easy for children to remember the numbers in the houses shown below:

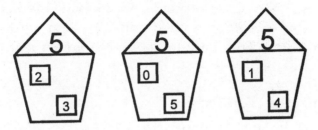

In contrast, they will have no difficulty at all remembering the numbers when they are arranged in sequential order like this:

Engage your students in a discussion about why this pattern is easier to remember than the previous one. See if they can discover on their own the sequence of numbers in the top floors: 0, 1, 2, which then continues from right to left in the bottom floor (3, 4, 5).

The Backwards C

A movement and shape that emerges in the number sequence is that of a backwards *C* (see illustration). This backwards *C* motion can provide a kinesthetic link to help the child recall the order in which the numbers are placed in each street of houses. Encourage the children to swing an arm in the backwards *C* motion as they look at their houses.

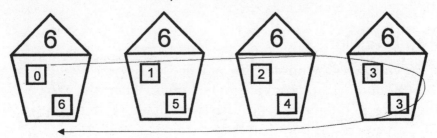

Exploration

Now that students have had an opportunity to discover for themselves the patterns in the number houses on the street you are targeting, they will move to pencil-and-paper addition and subtraction tasks using the number houses.

3. Discovery Worksheet

Give each child a copy of the version of worksheet 7.4 for the particular number you are working on ("Figure out which families can live in each house," pages 125–28). This worksheet mirrors the discovery activity the children have just completed. Now, however, they write the appropriate numbers in the house windows rather than using preprinted numbers. Remind them of the self-assessment guidelines: They have to use every number provided, and they cannot use any number twice.

4. Practice Houses Worksheet

Locate and copy the version of worksheet 7.5 (pages 129–33) that has the correct number of houses you need for the street you are on. Make enough copies for the class and trim along the top and bottom dotted lines to make equal thirds. After the children write their names on their papers, have them turn the papers face down and fold the bottom flap up first on the dotted line. Then have children fold the top section under. When they turn the folded paper over, they should be looking at only the top row of houses. Tell them what the attic number is, and have them write that number in every attic. Then have them write a zero in the top window of the first house. (You could also write these numbers in on a master copy of the worksheet before copying for the class, as illustrated below.) Now have students fill in the numbers to complete the families on the street. When they have completed the top row, students should turn their paper over and complete the center section, then finish by doing the bottom section. Repeat this activity until the children can dash off the numbers without difficulty.

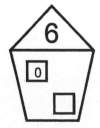   

5. All and No More

When the children have finished the practice houses worksheet, ask whether they have found all the possible combinations. When the class establishes that they have, point out that these few combinations are the only ways they will be able to divide that number. For example, for the target number six, they will make one house $0 + 6$, another $1 + 5$, then $2 + 4$ and lastly $3 + 3$, and those are the only four combinations they will be able to make that equal six.

6. My Two Hands

Do a quick review of the "my two hands" combinations for the target number (see page 65). At this point, encourage the children to make a sum on their hands (such as $2 + 4$), then to close their eyes and "see" their hands showing $2 + 4$. Thereby you are teaching them to draw upon imprinted pictures of the sums.

7. Practice Strips

Use the same practice house strips you used for step 4 (choose the appropriate version of worksheet 7.5, pages 129–33), but this time, before you copy enough for the class or group, fill in the upper window number on each house. Instead of following the established order for Stony Brook, write the numbers in a different order (see illustration). Explain that the families got tired of living in their apartments and decided to trade around on their street. The only rules in place now are that no two houses can be alike, and no more than the attic number can live in a house. I give students three sets of practice strips to complete, one at a time, so they cannot look at the problems they just completed to fill in the next ones.

  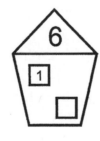 

8. Families Went Walking

Locate the version of worksheet 7.6 (pages 134–37) that matches the target number you are practicing. This series of worksheets has an attic and upper window number, leaving the lower floor to be filled in (which is the operation of subtraction). Referring to the numbers provided at the bottom of the page, students choose which family is missing from each house and write that number in the empty window. The children may wish to invent reasons why the missing families left. Maybe they went to the store, the park, or Grandma's house. Your hope at this stage is that children can draw upon their memory of which numbers go together in each house without counting. If they need to look at their hands to figure out the answers, you may wish to stop and review steps 6 and 7 above.

9. Additional Practice

Depending on students' fluency and comfort level, I will give either more practice with the problems they are hesitant about solving or move to a full page of mixed addition and subtraction problems. For students who are confidently solving the problems thus far, move to step 10. For students who need additional practice before moving on, you can make a copy of the relevent version of worksheet 7.5, write in appropriate problems, then photocopy enough for all the students who need practice.

10. Mixed Addition and Subtraction Practice

Worksheet series 7.7 (pages 138–41) contains both addition and subtraction problems for each target number. (Present only the worksheet for the number family you are targeting.) Each mixed practice sheet targets only one sum, but because there are various addition and subtraction possibilities for that number, it is easy to provide a page of problems. Here is how I describe the problems to the children:

> **Subtraction:** If the attic number is present and only one number is in the top part of the house, this means a family is missing (out for a walk, at the store, on a picnic, or whatever reason you come up with). The property manager has to figure out which family went walking.

> **Addition:** If the two families are there but the attic number is missing, a tornado wiped out the attic signs, and you, the property manager, must replace the signs immediately. The attic number has to equal the number of people in the bottom family plus the number of people in the top family.

With practice, the children will simply become used to seeing the number families together. Some children remember them geographically on the street as they are problem solving (visual mapping), some resort to looking at their two hands, and some make other connections.

11. Which House?

If you are planning to begin moving to traditional addition and subtraction problems at this point, the next step is to match addition and subtraction problems to the houses they come from (worksheet series 7.8, pages 142–47). (Present only the worksheet for the number family you are targeting.) This activity is optional but will help students understand the various operations that can be performed with just three numbers. For each problem, the children draw a line from the problem to the house containing the numbers that reflect that problem. For example, 2 + 4 would match a house that has six in the attic, two upstairs, and four downstairs. The child would draw a line to that house. Similarly, 6 − 2 would correspond to the same house.

The purpose of this activity is to show the children that a single combination of three numbers can yield up to four different problems, but that if they know one trio of numbers, they can easily solve all four problems.

12. Writing Problems

Worksheets 7.9a–h (pages 148–53), another optional series of worksheets, take the transition to symbols one step further, as children translate each house into traditional arithmetic problems. Present only the version of the worksheet for the targeted number family and work with the children as they need help in completing it. Discuss the various possibilities, reminding them that they can "lose" either attic numbers or families to make problems for each house. Discuss which symbol they would use for each type of situation. For the attic number six, and the house with two upstairs and four downstairs, they would be able to write two subtraction and two addition problems: $6-2$ (by losing the bottom family), $6-4$ (by losing the top family), $2+4$ (by losing the attic number), and $4+2$ (by losing the attic number). You might need to model completing this worksheet the first time you use it, but after that, the children should understand what they are doing for subsequent streets.

You have now completed the basic teaching procedure. If you have just worked through Third Street, go back to step 1 and start over with Fourth Street. Once you reach this point with Fourth Street, you can begin presenting practice sheets that have mixed problems for threes and fours, as described in step 13.

13. Mixed Numbers Practice

Once your class has mastered all the possible problems for Third and Fourth Streets individually, give them the mixed addition and subtraction practice sheets for threes and fours (worksheet 7.10a, page 154). Then, go back through steps 1–12 for Fifth Street, followed by Sixth Street. Once the children have mastered the Fifth and Sixth Street problems separately, you can present worksheet 7.10b (page 155), which has addition and subtraction problems for five and six, and worksheet 7.10c (page 156), which mixes problems for three, four, five, and six. Similarly worksheet 7.10d (page 157) has mixed problems for seven and eight; worksheet 7.10e (page 158) has mixed problems for nine and ten; and worksheet 7.10f (page 159) has mixed problems for seven, eight, nine, and ten. Present these worksheets as students master the targeted problems.

14. Additional Practice

Provide as little or as much practice with mixed problems as children need. You will find a page of empty houses that you can use to create individualized worksheets for extra practice (see worksheet 7.11, page 160). For example, you might present representative problems for all the numbers from three to ten to assess whether children have mastered them all.

You can target a variety of types of problems using worksheet 7.11. First make a photocopy of the worksheet and decide which number or numbers to target. Then decide which of the following type or types of problems you want students to practice:

- For discovery problems, fill in the attic numbers and let students write in both families.

- For subtraction problems, fill in an attic number and one family for each house, then let students write in the other family.

- For addition, fill in both families and let students fill in the attic numbers.

Optional Teaching Strategies

My rule of thumb is that students should be able to learn one street per practice session. If some of your students have difficulty learning certain streets, you will want to model making additional connections, so they learn to make connections on their own. *Remember, the foundation of this system is that connections drive learning.* Here are some connections between symbol and prior knowledge that have worked for students who had difficulty remembering the streets.

Label the Houses

I give associative names to the number combinations that recur on every street (see illustration on the facing page). These four house types (Tornado House, Counting House, Parents' House, and Split House) account for every combination on Third through Seventh Streets.

Hand And . . . Houses

Any house that has a five on a floor qualifies as a Hand And . . . House. Of course, a Hand And . . . House will at the same time be one of the four previously described house types (such as a Parent House—2 + 5—or a Split House—4 + 5). However, identifying the Hand And . . . Houses will help to tie those number combinations to the child's own hands (which are very familiar), as well as to the five-frame format that will be used for higher levels of math. The black houses in the illustration are the Hand And . . . Houses.

Stony Brook Village

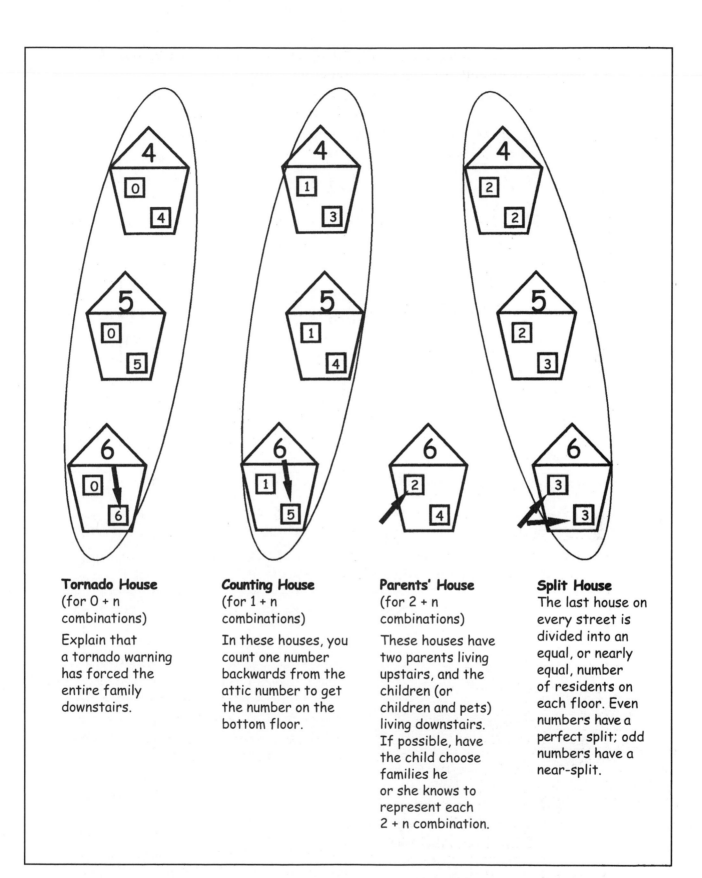

Tornado House
(for 0 + n combinations)

Explain that a tornado warning has forced the entire family downstairs.

Counting House
(for 1 + n combinations)

In these houses, you count one number backwards from the attic number to get the number on the bottom floor.

Parents' House
(for 2 + n combinations)

These houses have two parents living upstairs, and the children (or children and pets) living downstairs. If possible, have the child choose families he or she knows to represent each 2 + n combination.

Split House

The last house on every street is divided into an equal, or nearly equal, number of residents on each floor. Even numbers have a perfect split; odd numbers have a near-split.

Notice that on each successive street, the Hand And . . . House moves one place to the right. Connect these houses purposefully with the child's hands, and use the "my two hands" language as you model the combinations on your hands:

- Five is a hand and no more.
- Six is a hand and one more.
- Seven is a hand and two more.
- Eight is a hand and three more.
- Nine is a hand and four more.
- Ten is two hands.

Make Up a Story for Each Street

In order to create solid connections for learning, have the children choose a family name for each street. For example, whichever street corresponds to the number of people in the child's own family might be named with that child's surname, and other streets could be named for other families well known to the child. Assume, for example, that Seventh Street is the Browns' street. Here is their story:

The Brown family has two parents, four kids, and a dog named Goliath that is very antsy.

Tornado House: Everyone is in the basement.

Counting House: Goliath runs upstairs; the rest of the family stays downstairs.

Parents' House: The parents go upstairs to make sure it's all clear, telling the kids to wait downstairs. Goliath promptly runs downstairs again.

Split House: Having checked on the kids downstairs, Goliath runs back upstairs to be with the parents.

Have the child make up a story like this for every street with which he or she has difficulty. I cannot emphasize enough how important it is that the child should invent the family and the story, because the story is intended to provide a meaningful link to the number facts. A story you make up might have associations that are useful for you but not for the child.

Strategies for Learning Eighth Street: The Label Street

Notice that Eighth Street contains one of each kind of house. It reminds me of those people who buy only clothes with designer labels. Take a look:

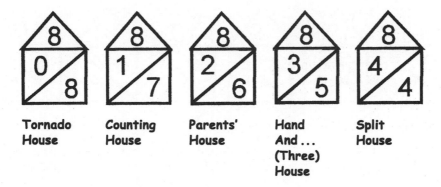

Tornado House   **Counting House**   **Parents' House**   **Hand And ... (Three) House**   **Split House**

Strategies for Learning Ninth Street

The one house on Ninth Street that cannot be described using the existing labels (Tornado House, Counting House, Parents' House, and Split House) has a unique feature that makes recall simple. Notice in the illustration that there is a house with all curved numbers. The attic number (nine) is round at the top, the six is rounded at the bottom, and the three has two rounded sides. This contrasts with the Slide Numbers House (also Parents' House) next to it, containing a two and a seven, both of which slide or slant from upper right to lower left.

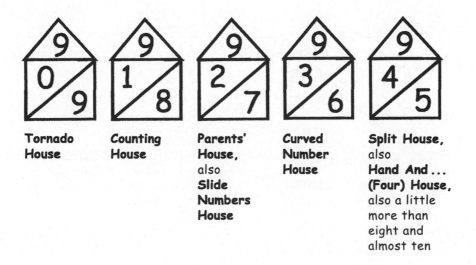

Tornado House   **Counting House**   **Parents' House,** also **Slide Numbers House**   **Curved Number House**   **Split House,** also **Hand And ... (Four) House,** also a little more than eight and almost ten

Strategies for Learning Tenth Street: The Longest Street

Begin by labeling the recurring house names, then tie in the remaining houses to the labeled houses or their relationship to ten:

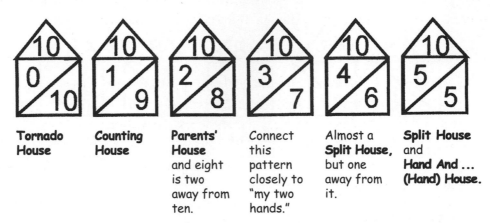

| Tornado House | Counting House | Parents' House and eight is two away from ten. | Connect this pattern closely to "my two hands." | Almost a Split House, but one away from it. | Split House and Hand And ... (Hand) House. |

It also helps to picture these combinations using "my two hands" (see illustration). How many away from ten is six? (Four) What about seven? (Three) And eight? (Two) For the 3 + 7 house, "my two hands" is the strongest connection I have found. Show your left hand with all five fingers together and the fingers of the right hand doing a two-three split. Position the two right-hand fingers right next to the fingers of the left hand, to make seven. Do the same for the 2 + 8 house, with all five fingers of the left hand and a three-two split on the right-hand fingers. Allow plenty of practice.

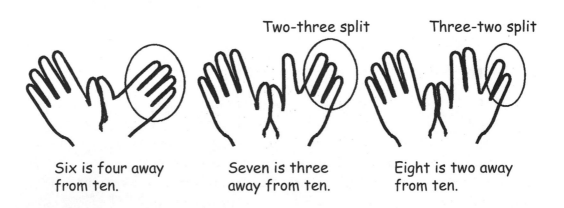

Six is four away from ten. Seven is three away from ten. Eight is two away from ten.

Ask the Child

One of your most powerful tools for making connections will be to ask students repeatedly, "How can you remember this?" Doing this will prompt them to create mnemonic aids for learning and will give them the sense of being in control of their own learning. The mirror image to that practice is to ask repeatedly, "How did you remember that?" At first children might simply shrug and say, "I just did." Or "I don't know." But persist in asking. Before very long, they will begin to consider how they remembered, thereby rapidly developing powerful metacognitive habits that will strengthen their learning process. They will move from passive bystander to proactive learner.

Extension Activities

For those children who want an extra challenge, I have included samples of extra practice worksheets in Appendix A (worksheets 7.12a–f, pages 161–66). Some have story problems, and others are blank so students can invent their own problems. At this stage, encourage the children to be as creative as possible in their work. If they are able to put their own unique stamp on the work, they will take pride in it.

House Flash Cards

A set of "house" flash cards can be found at the end of Appendix A (worksheets 7.13a–f pages 167–72). These have many possible uses:

- Line them up in a pocket chart, one street per row, to make a visual for Stony Brook Village.
- Mix them up and have students put them in the correct order.
- Use for practice with a partner: One child covers up a number with his or her thumb, and the other child identifies the covered number.
- Punch a hole in each card and string them on a ring for a child to refer to during computation (although I find using "my two hands" more effective).

Informal Assessment Strategies

Observe the students as they work to determine whether any need further practice with a single target (attic) number before moving on to mixed problems. Usually, though, the children do not get seriously confused. If they do, this only brings up a teaching moment. As you discover which problems cause difficulties, try strategies that relate the problem to its family, rather than expecting the child to memorize an isolated fact.

The mixed numbers practice pages (worksheet series 7.10, pages 154–59) can also serve as a first assessment, if desired. Remember to use these worksheets to determine which problems need further teaching and which might need stronger connections. Remind the children, if necessary, that if the attic number is missing, they have to replace the missing number quickly. They, of course, will remember that the attic number is there to tell everyone how many people can live in that house. If a floor is empty, that means one of the families went walking, and they are to determine which one. Worksheet 7.11 (page 160), the blank practice houses worksheet, is also useful for making up a page of problems for assessment purposes.

Conclusion

Enjoy yourself. Encourage your students to decide now and then what math work they would like to do that day. I suspect you will find that they ask for specific streets on which they need practice. I would suggest keeping cubbies in the room with prepared sheets for each type of problem. On "free math day," students can simply choose whichever sheets they want to review. Allowing students to choose the math work they do is very empowering for them. They will delight in making decisions for themselves within the parameters you set for them.

Appendix A
Reproducible Blackline Masters

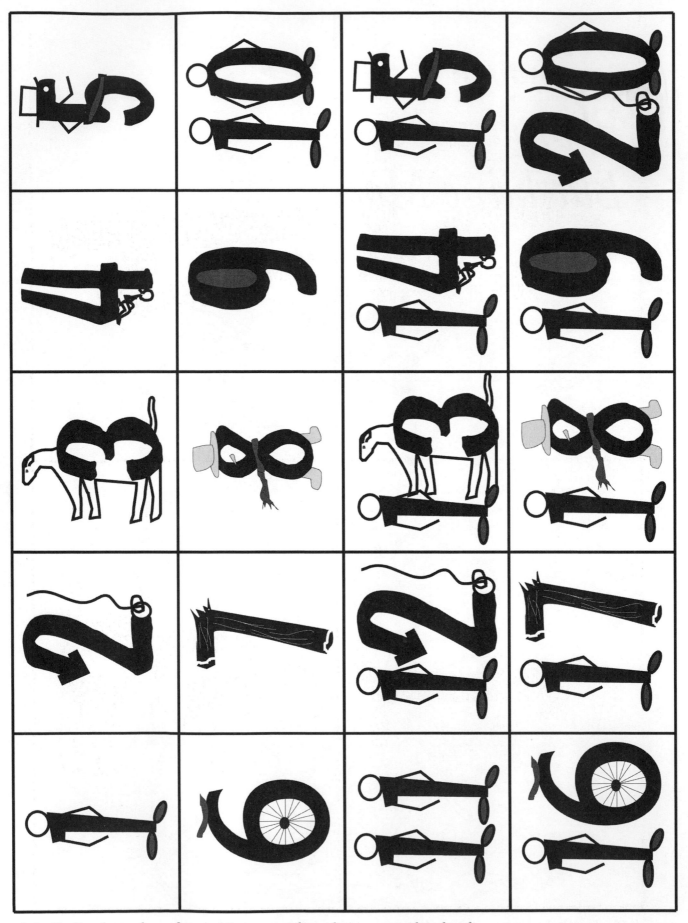

4.1. Stylized numbers for making a number chart or number book

4.2a. Stylized numbers for making a number chart or number book

4.2b. Stylized numbers for making a number chart or number book

 Addition and Subtraction, © 2002 Zephyr Press, Chicago • www.zephyrpress.com

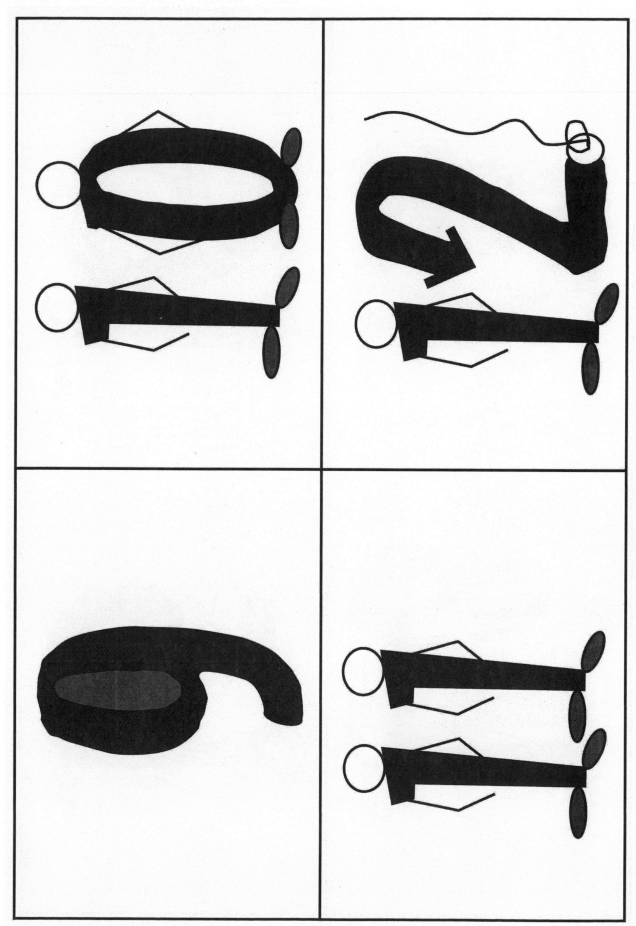

4.2c. Stylized numbers for making a number chart or number book

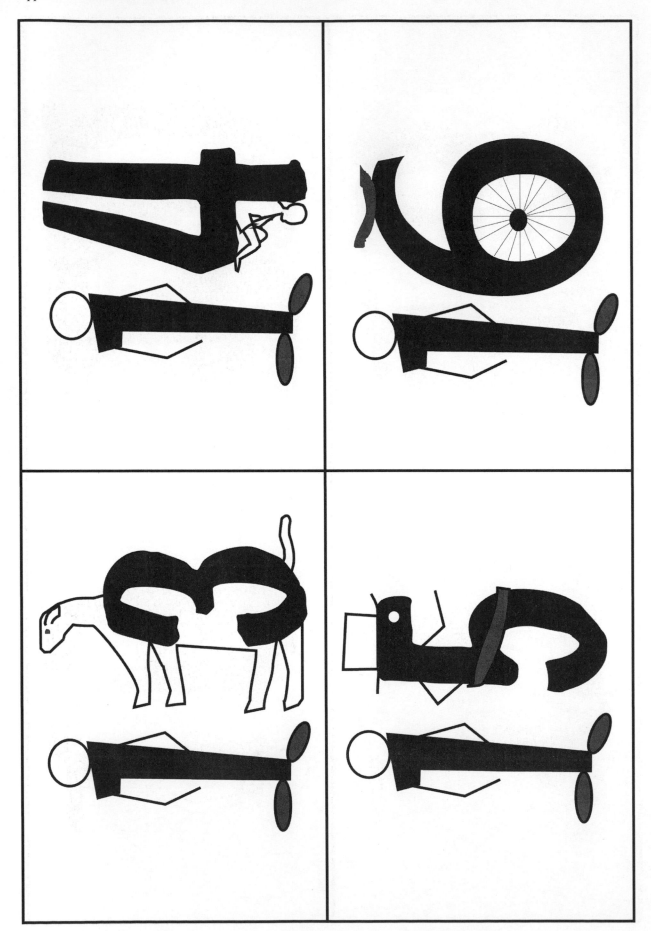

4.2d. Stylized numbers for making a number chart or number book

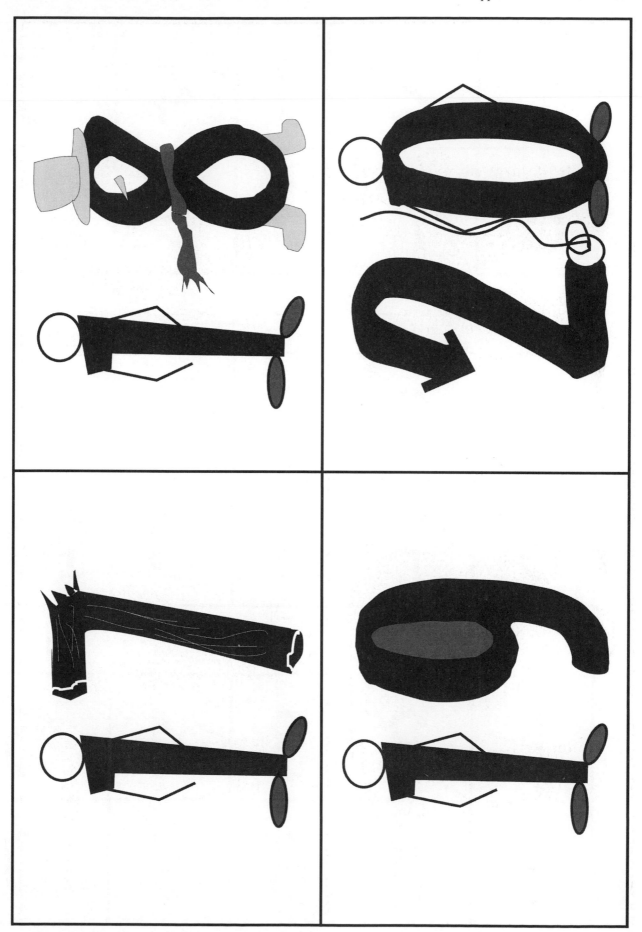

4.2e. Stylized numbers for making a number chart or number book

Name: _____

1	2	3	4	5
6	7	8	9	10

4.3. Fives chart to ten

- -

Name: _____

1	2	3	4	5
6	7	8	9	10

4.3. Fives chart to ten

- -

Name: _____

1	2	3	4	5
6	7	8	9	10

4.3. Fives chart to ten

Name: _____

4.4a. Ordering numbers to ten

9   7   2   4   10

5   1   3   6   8

- -

Name: _____

4.4a. Ordering numbers to ten

9   7   2   4   10

5   1   3   6   8

Name: _____

4.4b. Ordering numbers to 20

8	15	6	5	3
11	17	12	14	4
20	19	9	2	10
1	7	13	18	16

Addition and Subtraction, © 2002 Zephyr Press, Chicago • www.zephyrpress.com

1	2		4	
	7	8		10

Name: _____

4.5a. Missing numbers to ten—page 1

- -

		3		
6	7		9	10

Name: _____

4.5a. Missing numbers to ten—page 2

- -

1	2		4	5
		8		

Name: _____

4.5a. Missing numbers to ten—page 3

1		3		5
	7		9	
	12		14	
16		18		20

Name: _____

4.5b. Missing numbers to 20—page 1

- -

	2		4	
6		8		10
11		13		15
	17		19	

Name: _____

4.5b. Missing numbers to 20—page 2

1	2	3	4	5
6	7	8	9	10
11	12	13	14	15
16	17	18	19	20

Name: _____

4.6a. Five-frame chart to 20

- -

Name: _____

4.6b. Blank five-frame chart for practice

Name: _____

4.7. Blank five-frame chart for writing numbers

- -

Name: _____

4.7. Blank five-frame chart for writing numbers

- -

Name: _____

4.7. Blank five-frame chart for writing numbers

5.1. Blank dot card grid

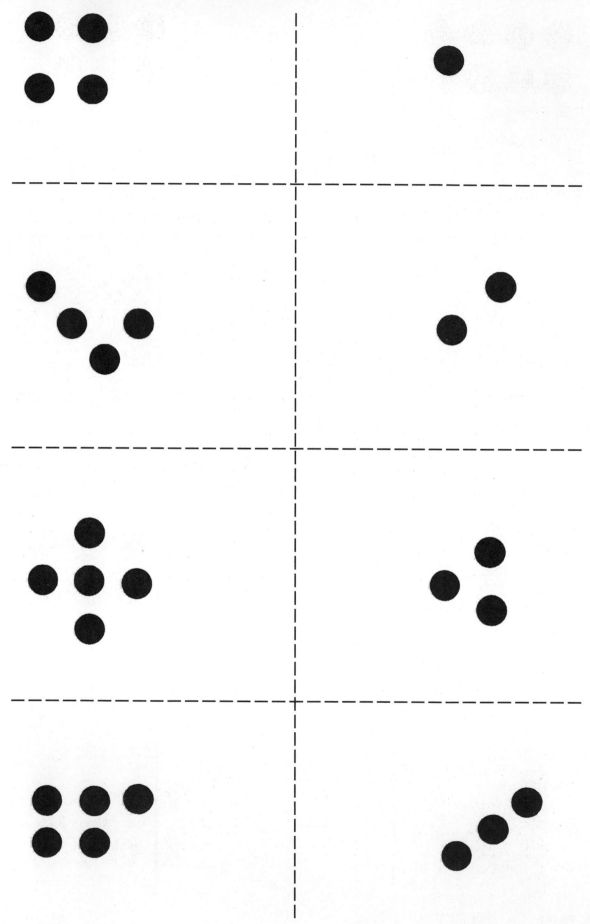

5.2a. Dot cards—page 1

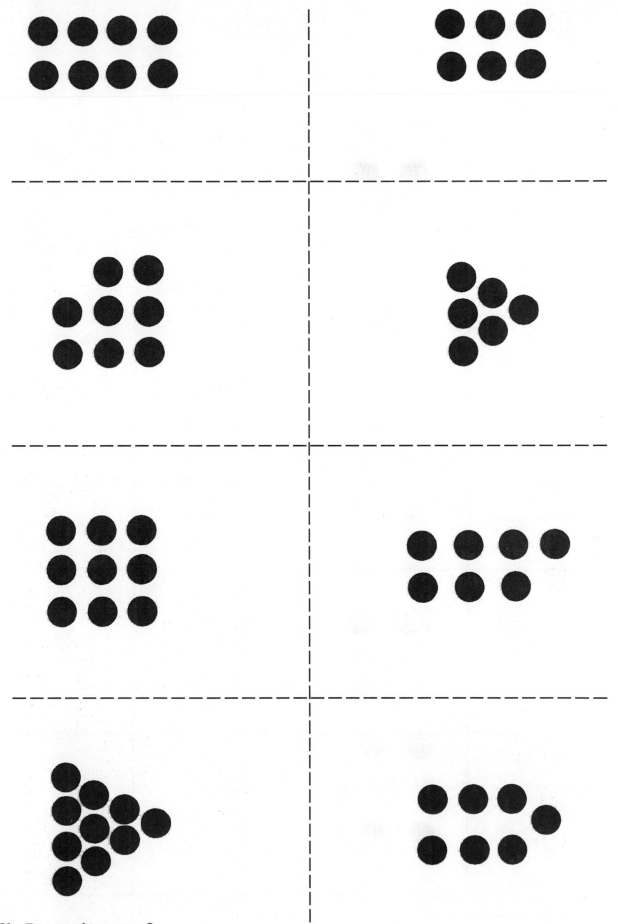

5.2b. Dot cards—page 2

Bingo

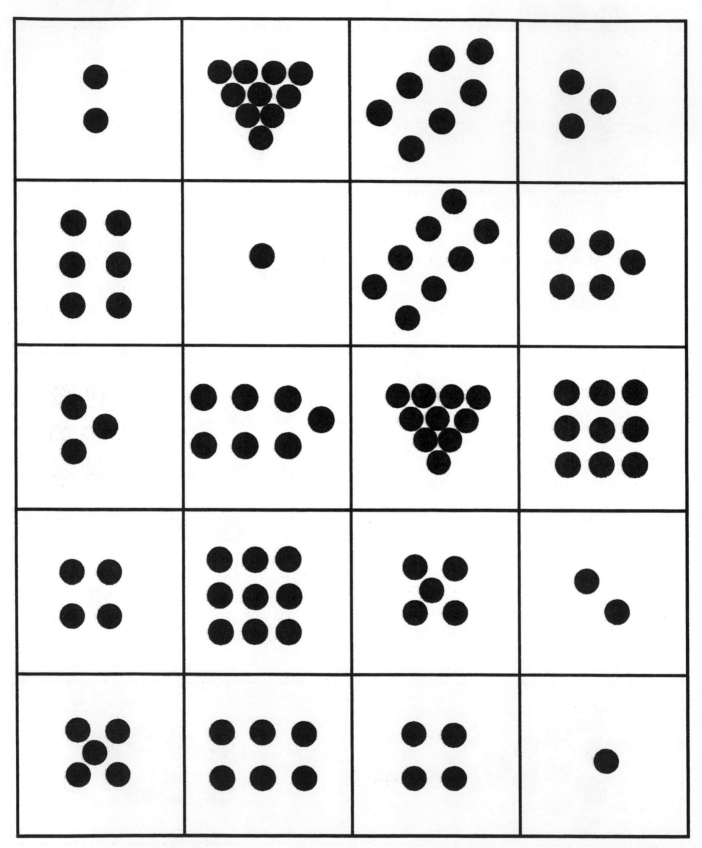

5.3a. Dot card bingo board 1

Addition and Subtraction, © 2002 Zephyr Press, Chicago • www.zephyrpress.com

Bingo

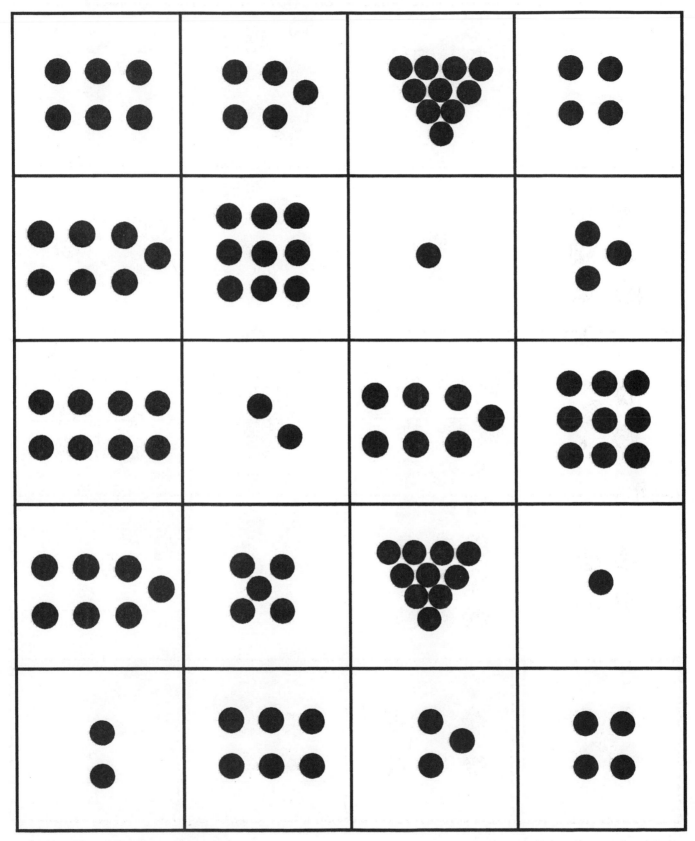

5.3b. Dot card bingo board 2

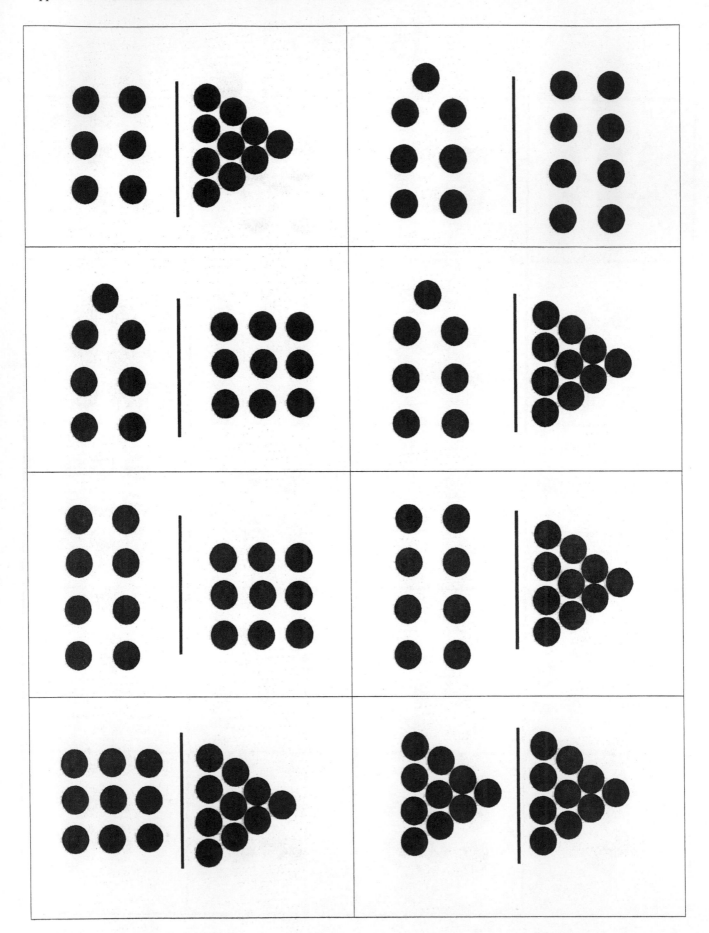

5.4a. Floor dominoes—page 1

Addition and Subtraction, © 2002 Zephyr Press, Chicago • www.zephyrpress.com

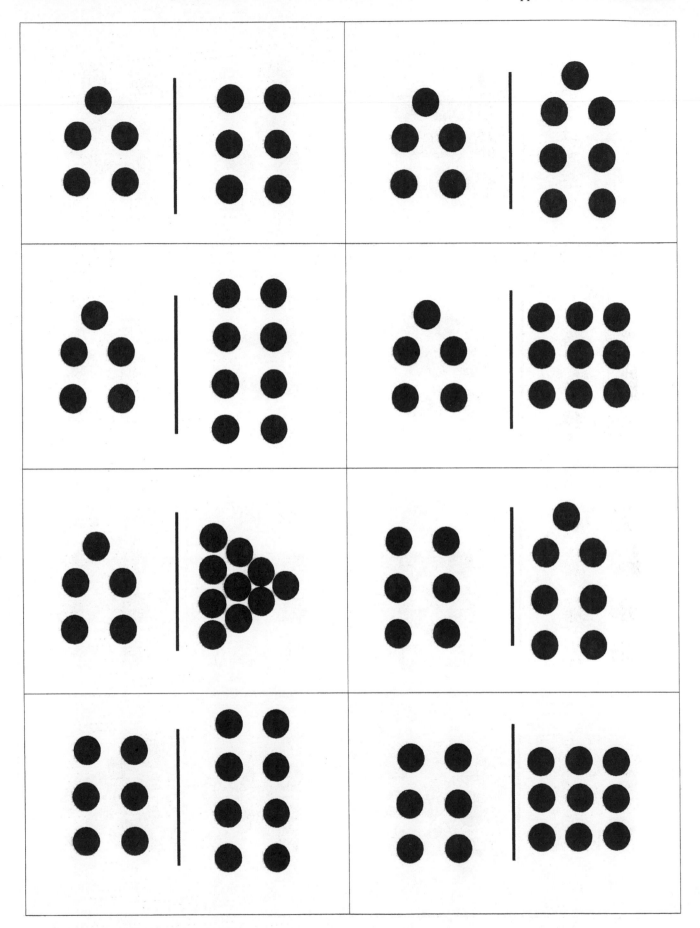

5.4b. Floor dominoes—page 2

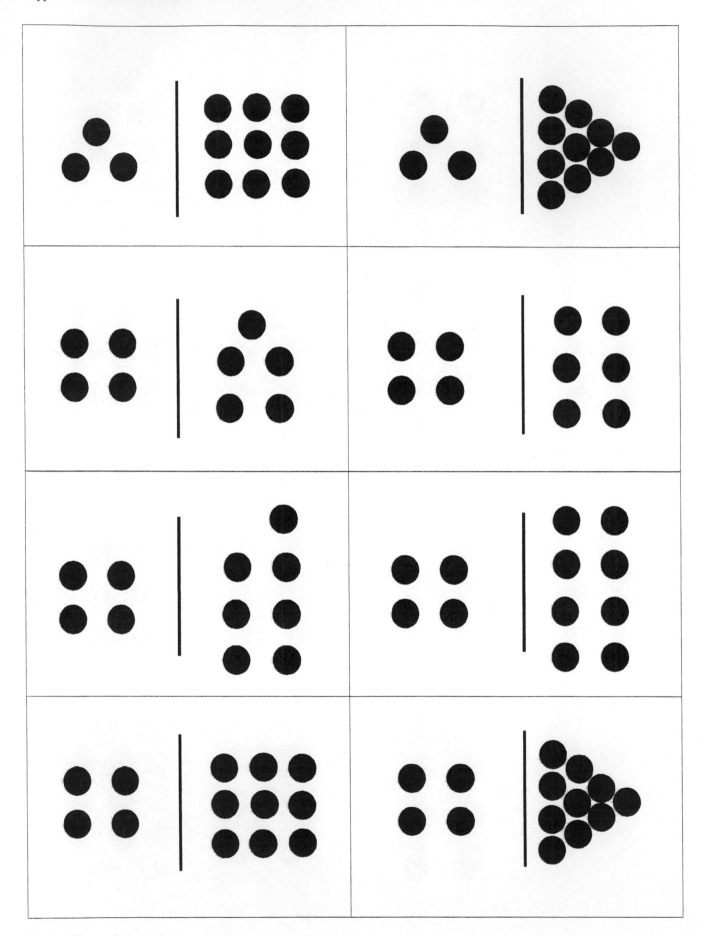

5.4c. Floor dominoes—page 3

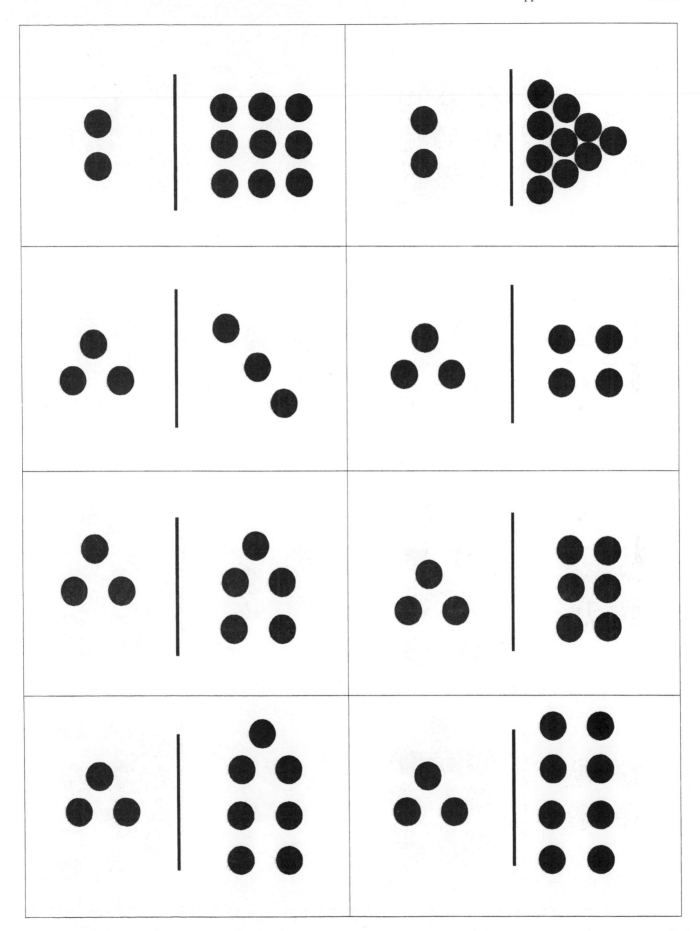

5.4d. Floor dominoes—page 4

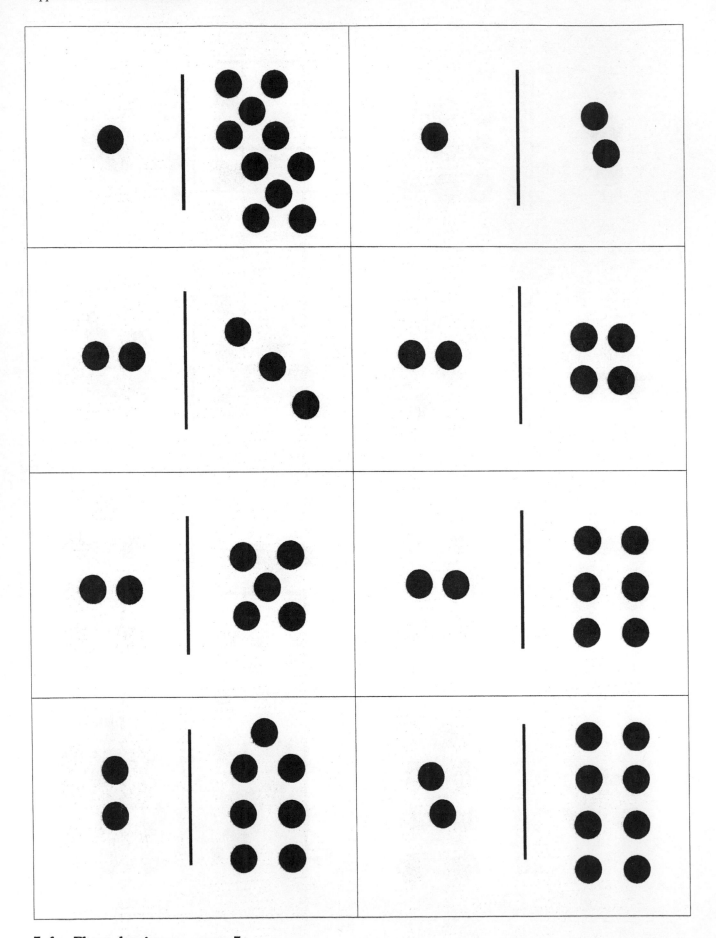

5.4e. Floor dominoes—page 5

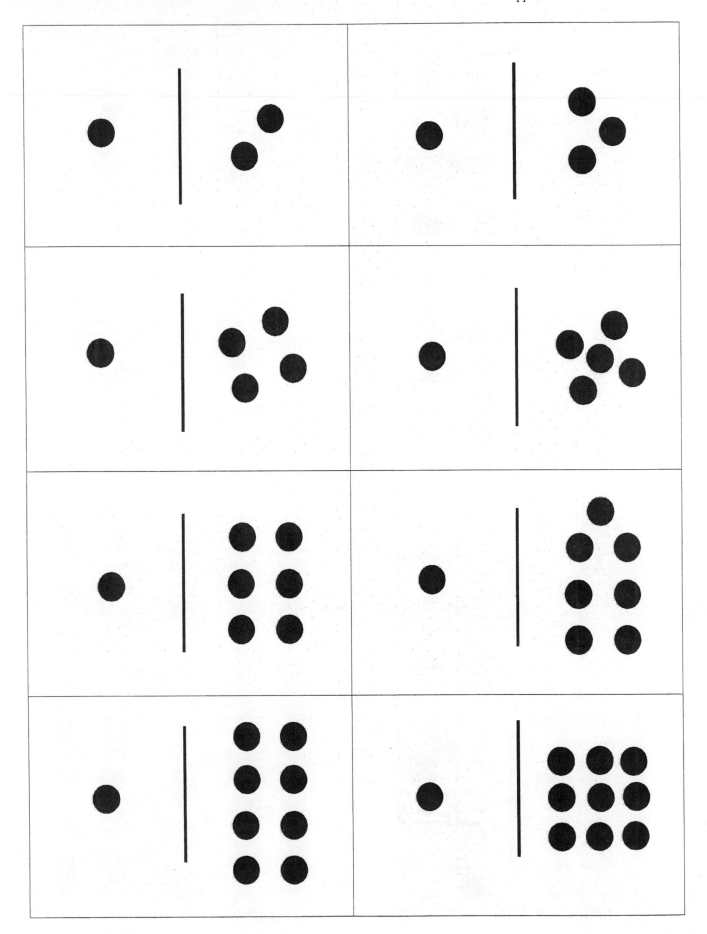

5.4f. Floor dominoes—page 6

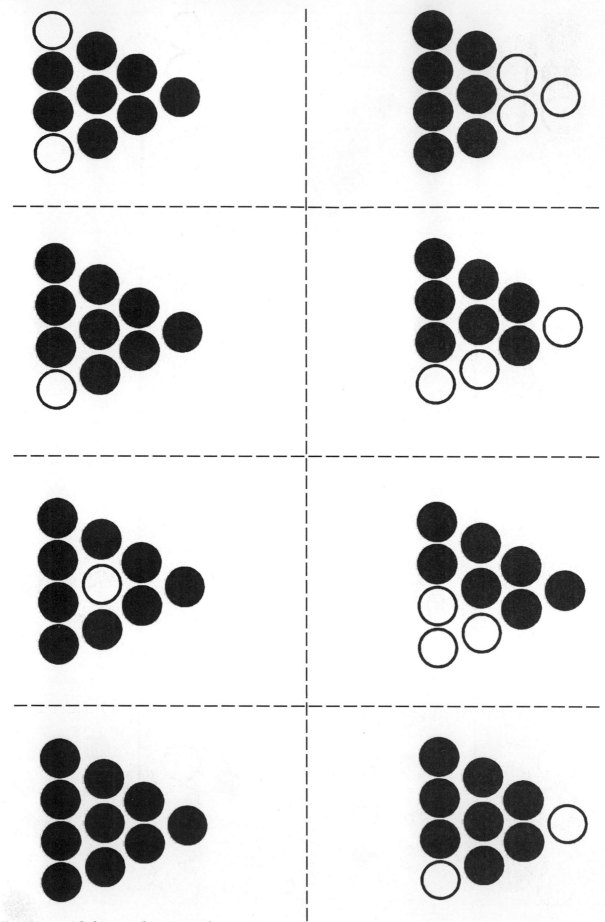

5.5a. Pyramid dot cards—page 1

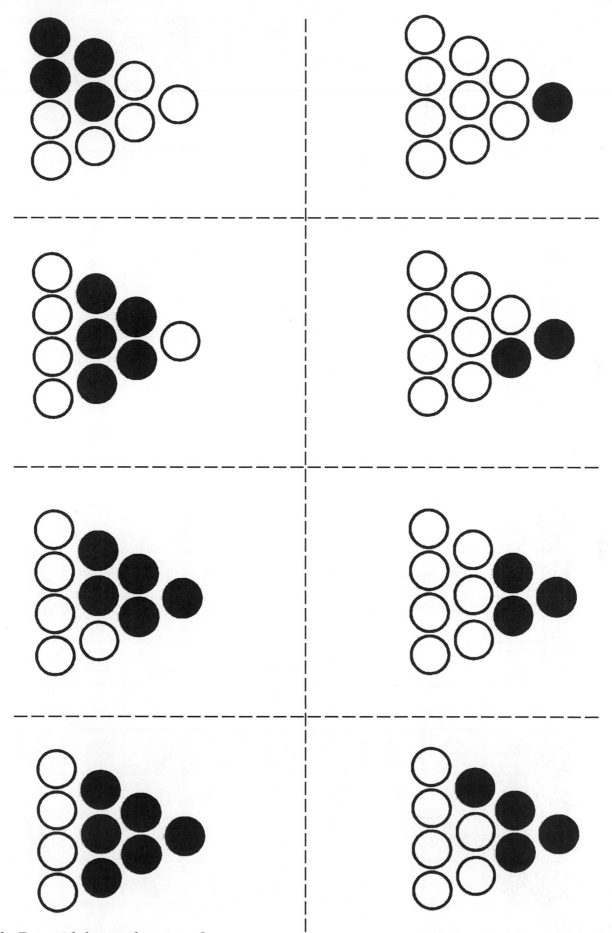

5.5b. Pyramid dot cards—page 2

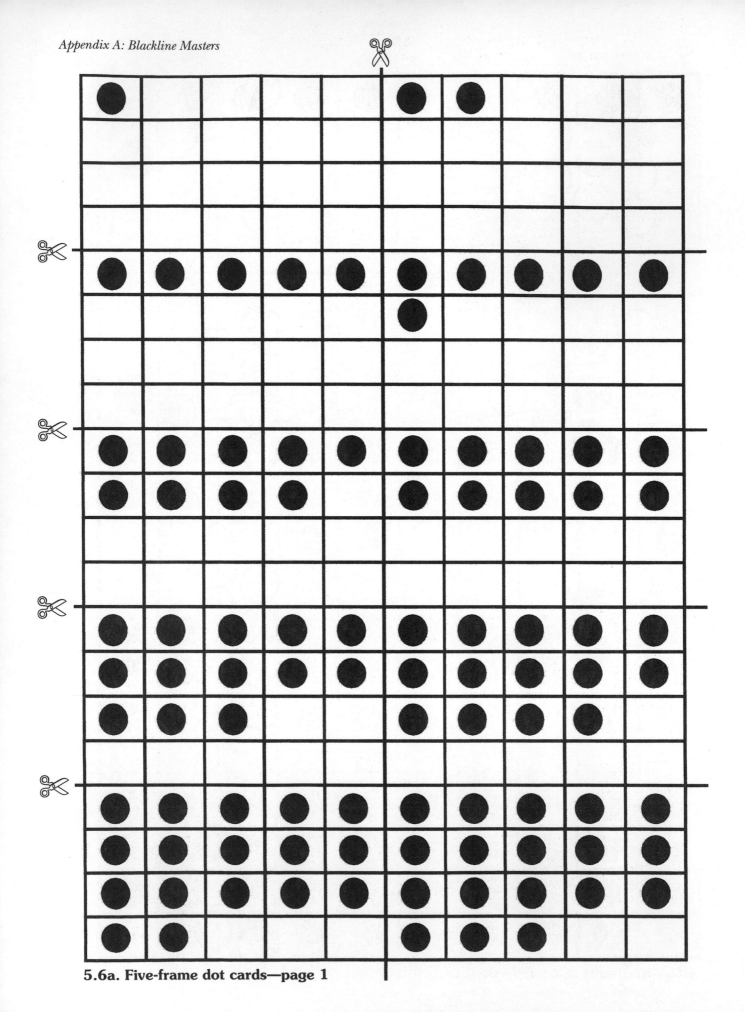

5.6a. Five-frame dot cards—page 1

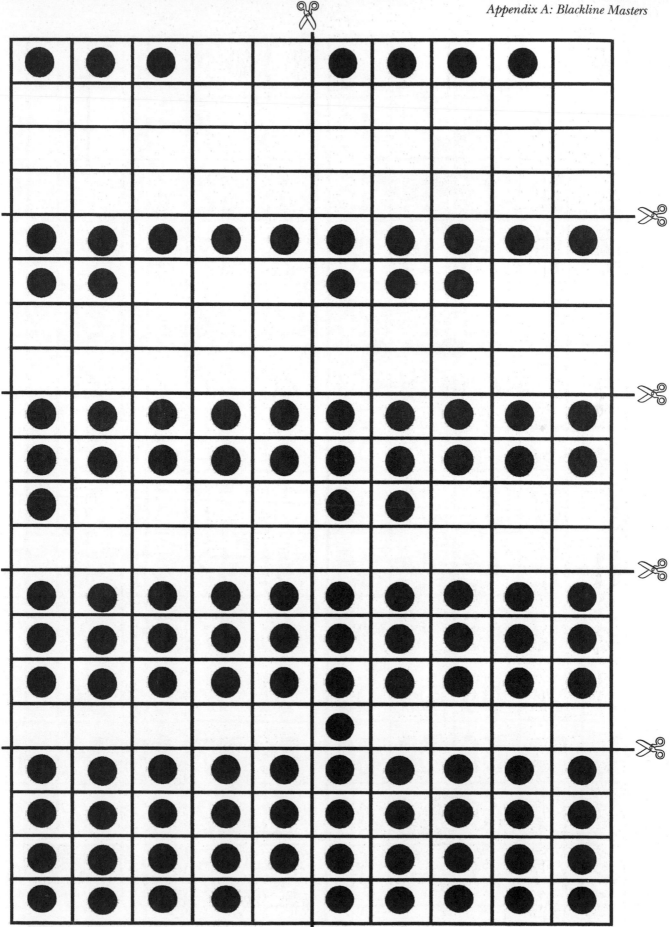

5.6b. Five-frame dot cards—page 2

1	2	3	4	5
6	7	8	9	10
11	12	13	14	15
16	17	18	19	20
21	22	23	24	25
26	27	28	29	30
31	32	33	34	35
36	37	38	39	40
41	42	43	44	45
46	47	48	49	50
51	52	53	54	55
56	57	58	59	60
61	62	63	64	65
66	67	68	69	70

1	2	3	4	5
6	7	8	9	10
11	12	13	14	15
16	17	18	19	20
21	22	23	24	25
26	27	28	29	30
31	32	33	34	35
36	37	38	39	40
41	42	43	44	45
46	47	48	49	50
51	52	53	54	55
56	57	58	59	60
61	62	63	64	65
66	67	68	69	70

5.7. Five-frame chart to 70

Addition and Subtraction, © 2002 Zephyr Press, Chicago • www.zephyrpress.com

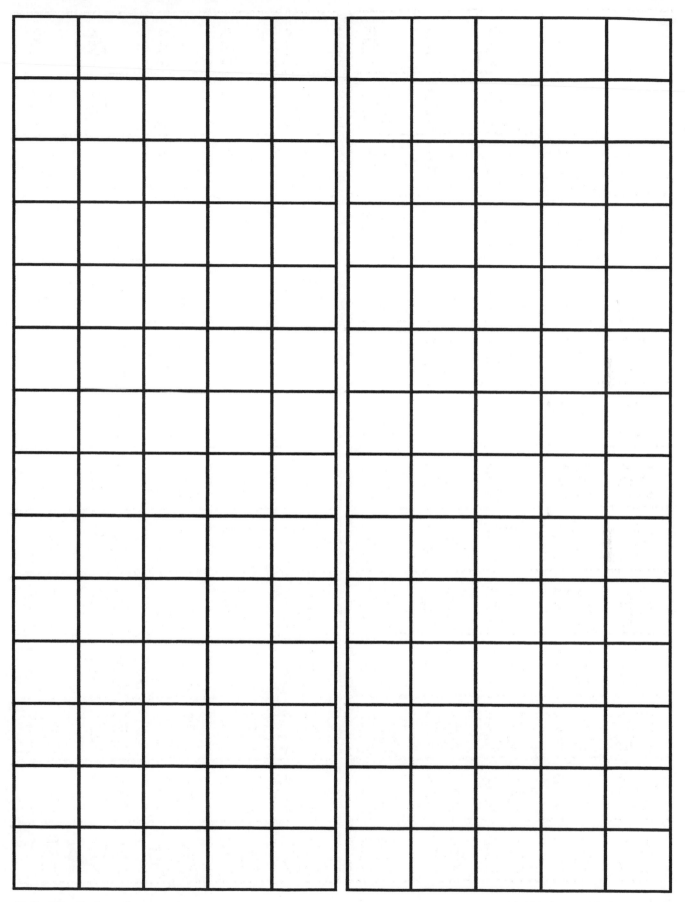

5.8. Blank five-frame chart

When You Say…
I See…

by _____

My little book of numbers

cut

fold

When You Say…
I See…

by _____

My little book of numbers

My Dot Critter Book

by _____

My second little book of numbers

---

cut

fold

My Dot Critter Book

by _____

My second little book of numbers

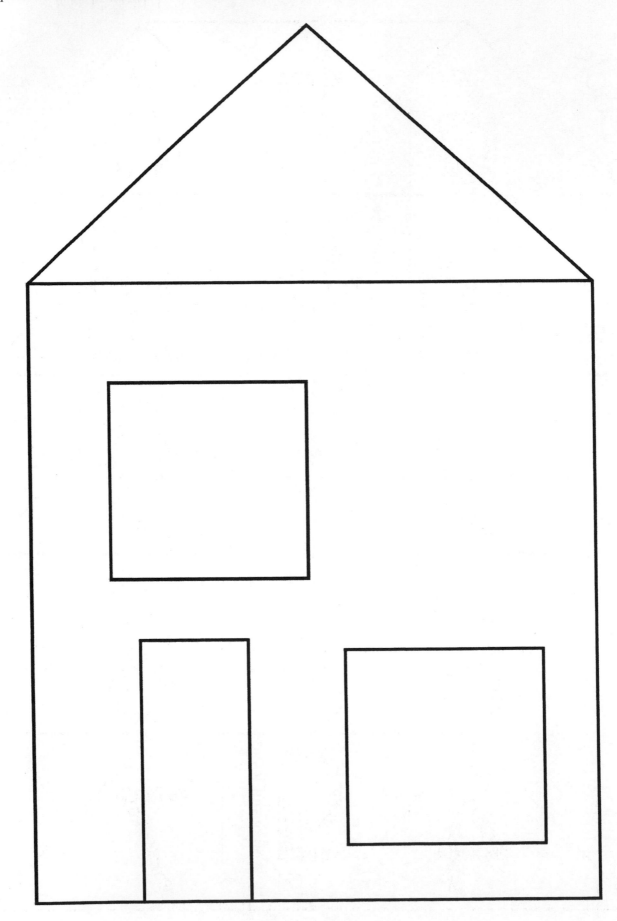

7.1. Felt-house pattern

7.2a Attic numbers—page 1

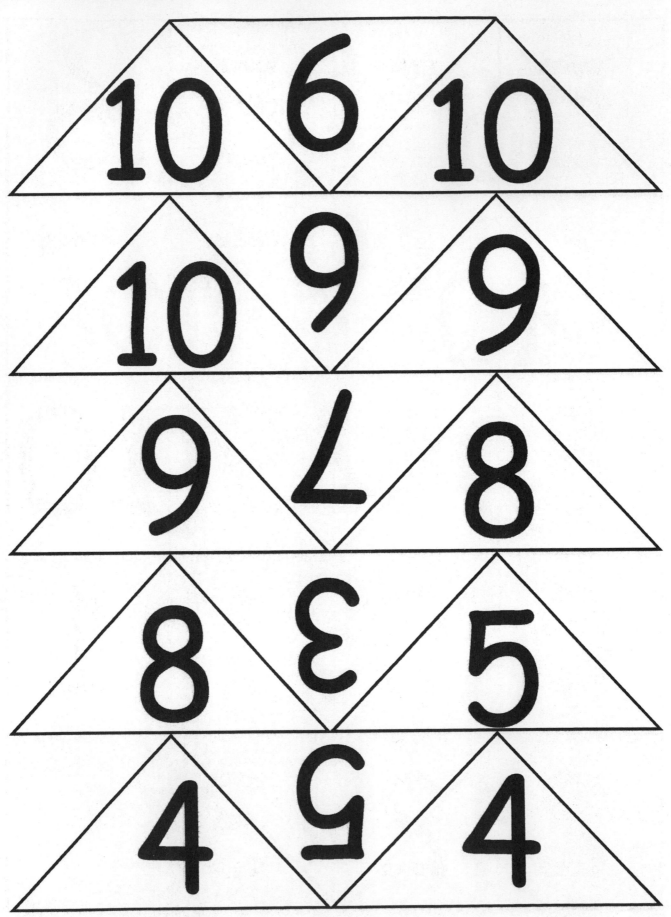

7.2b. Attic numbers—page 2

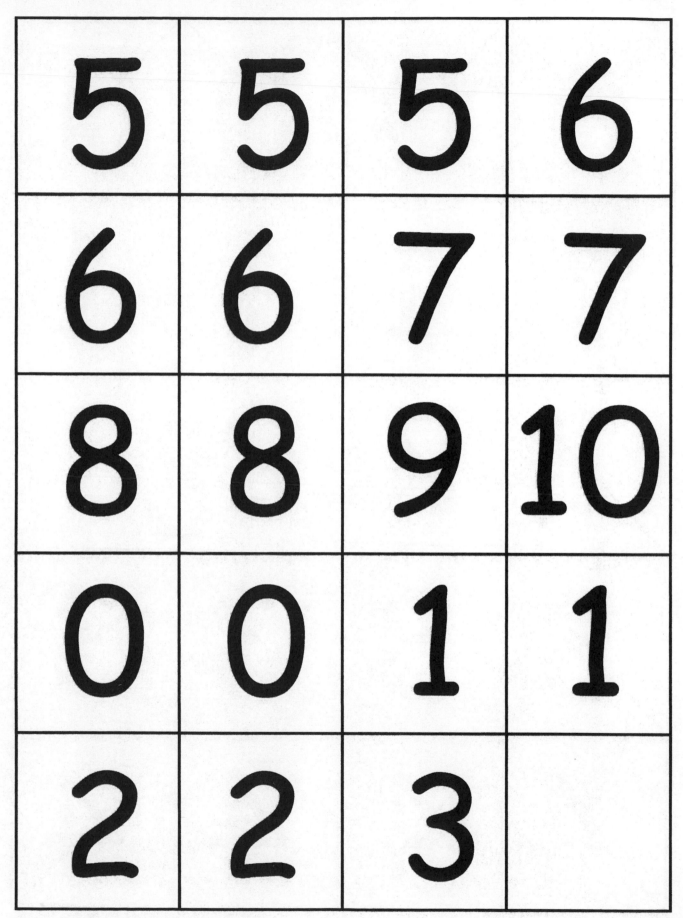

7.3a. Window numbers—page 1

0	0	0	0
1	1	1	1
2	2	2	2
3	3	3	3
4	4	4	4

7.3b. Window numbers—page 2

Addition and Subtraction, © 2002 Zephyr Press, Chicago • www.zephyrpress.com

7.4a. Threes

Figure out which families can live in each house.

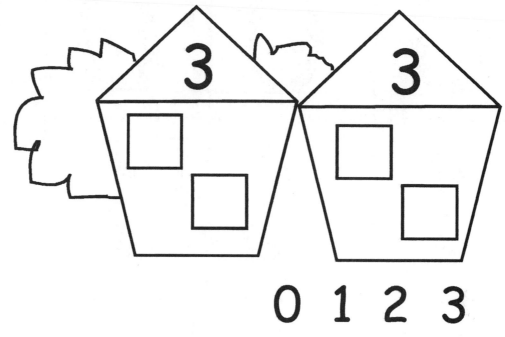

0 1 2 3

Name:_____

. . . .

7.4b. Fours

Figure out which families can live in each house.

0 1 2 2 3 4

Name:_____

7.4c. Fives

Figure out which families can live in each house.

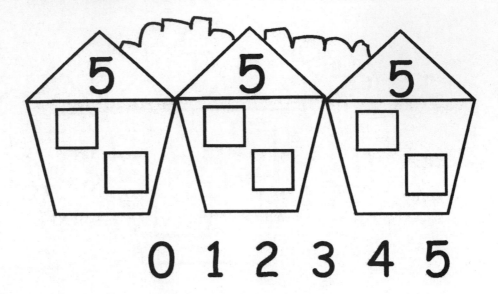

0 1 2 3 4 5

Name:_____

. . . .

7.4d. Sixes

Figure out which families can live in each house.

0 1 2 3 3 4 5 6

Name:_____

7.4e. Sevens

Figure out which families can live in each house.

0 1 2 3 4 5 6 7

Name: _____

. . . .

7.4f. Eights

Figure out which families can live in each house.

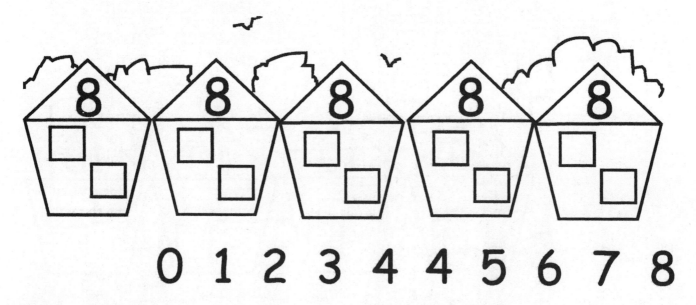

0 1 2 3 4 4 5 6 7 8

Name: _____

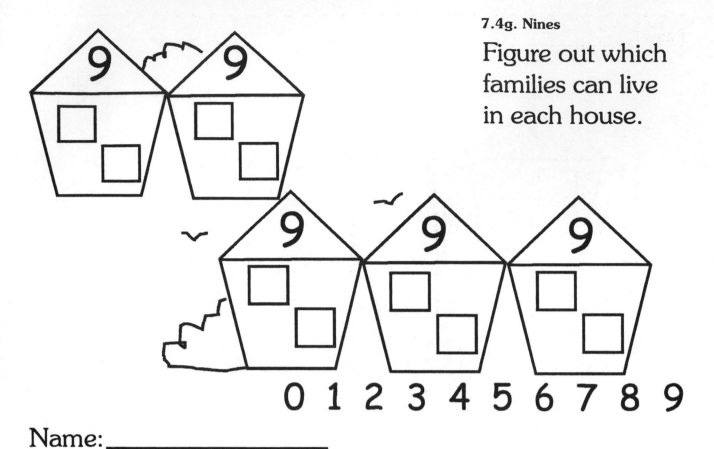

7.4g. Nines

Figure out which families can live in each house.

0 1 2 3 4 5 6 7 8 9

Name: _____

7.4h. Tens

Figure out which families can live in each house.

0 1 2 3 4 5 5 6 7 8 9 10

Name: _____

Addition and Subtraction, © 2002 Zephyr Press, Chicago • www.zephyrpress.com

Name_____

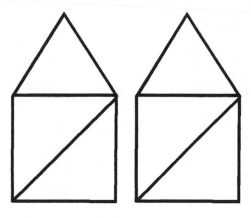

fold

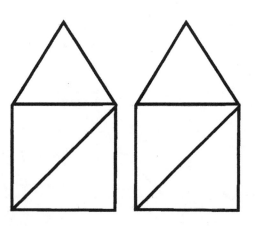

fold

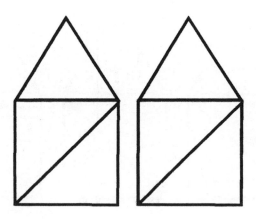

7.5a. Third Street practice houses

Name _____

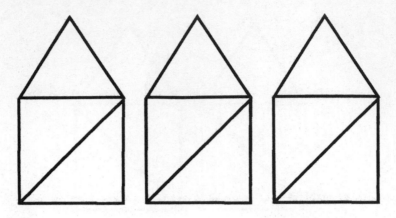

fold

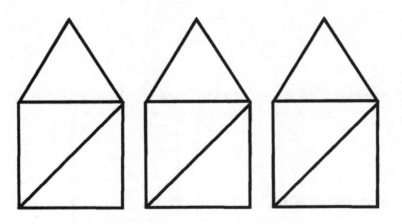

fold

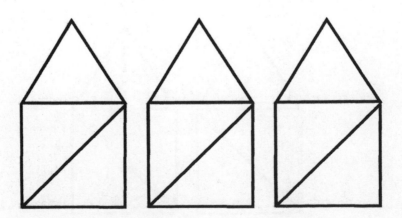

7.5b. Fourth and Fifth Street practice houses

Name_____

fold

fold

7.5c. Sixth and Seventh Street practice houses

Name_____

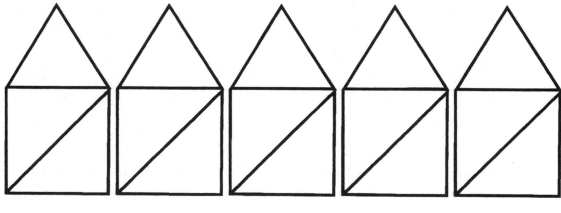

fold

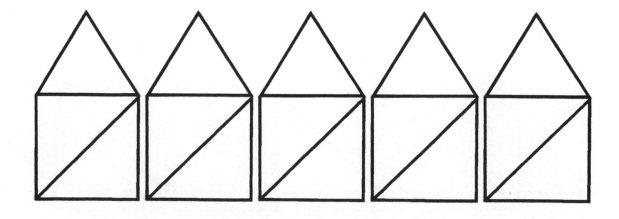

fold

7.5d. Eighth and Ninth Street practice houses

Name_____

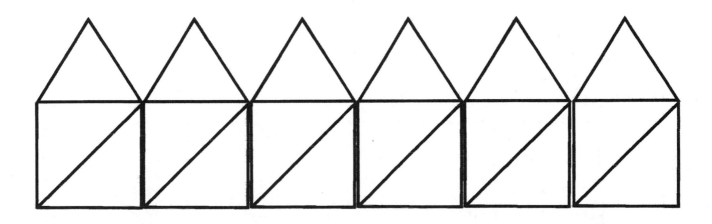

fold

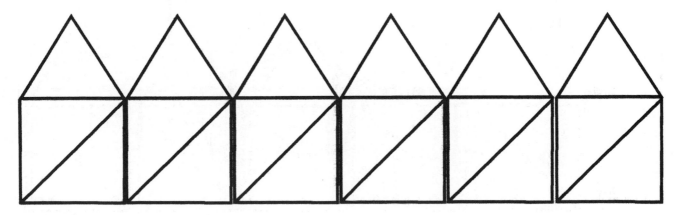

fold

7.5e. Tenth Street practice houses

7.6a. Subtraction problems for three

Please help me figure out which families went walking.

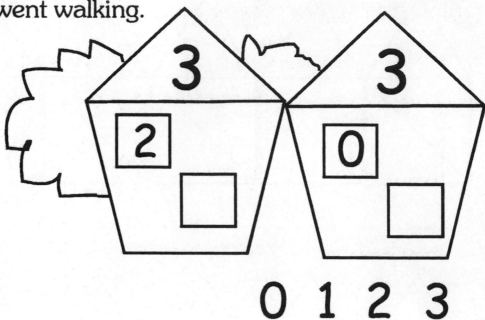

Name: _____

. . . .

7.6b. Subtraction problems for four

Please help me figure out which families went walking.

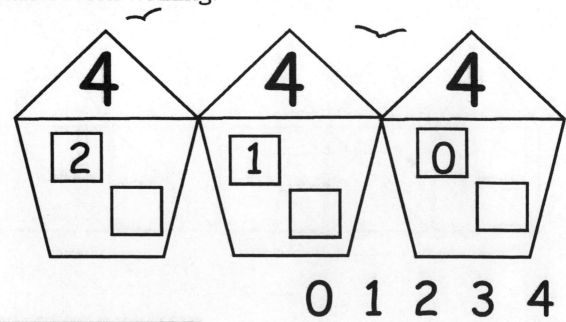

Name: _____

    Addition and Subtraction, © 2002 Zephyr Press, Chicago • www.zephyrpress.com

7.6c. Subtraction problems for five

Please help me figure out which families went walking.

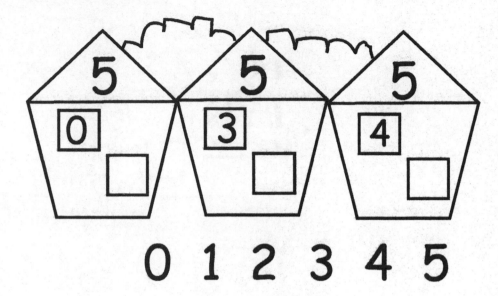

0 1 2 3 4 5

Name:_____

· ·

7.6d. Subtraction problems for six

Please help me figure out which families went walking.

0 1 2 3 4 5 6

Name:_____

7.6e. Subtraction problems for seven

Please help me figure out which families went walking.

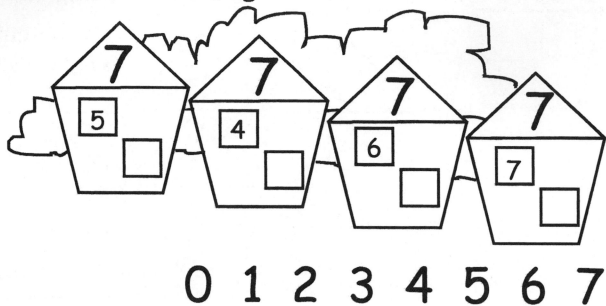

0 1 2 3 4 5 6 7

Name:_____

- -

7.6f. Subtraction problems for eight

Please help me figure out which families went walking.

0 1 2 3 4 5 6 7 8

Name:_____

7.6g. Subtraction problems for nine

Please help me figure out which families went walking.

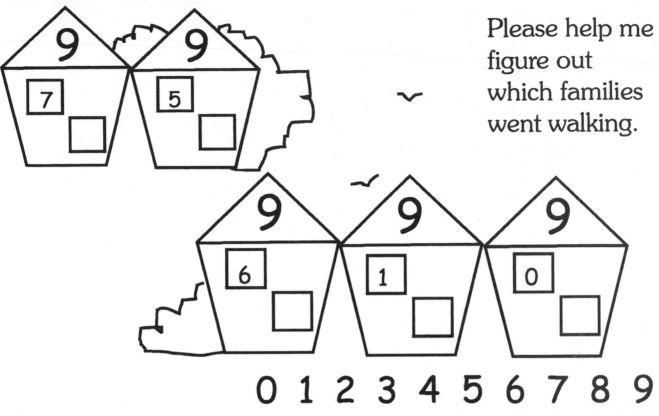

Name: _____

- -

7.6h. Subtraction problems for ten

Please help me figure out which families went walking.

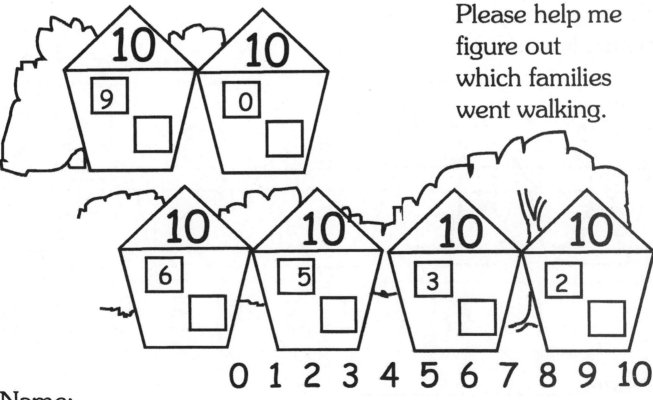

Name: _____

7.7a. Practice for threes   Name:_____

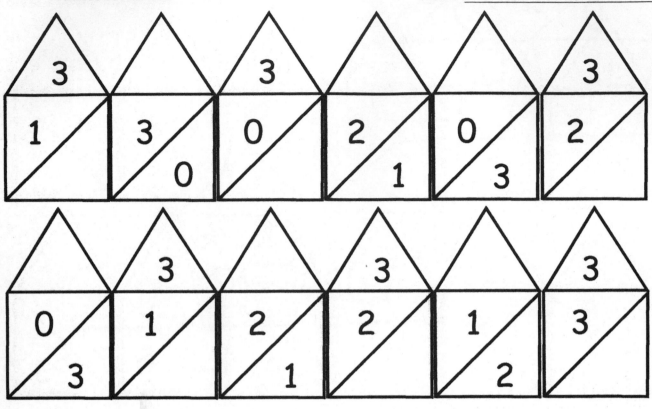

- -

7.7b. Practice for fours   Name:_____

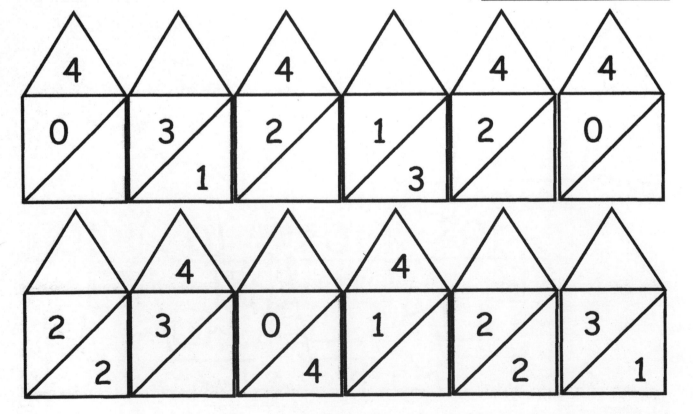

7.7c. Practice for fives

Name:_____

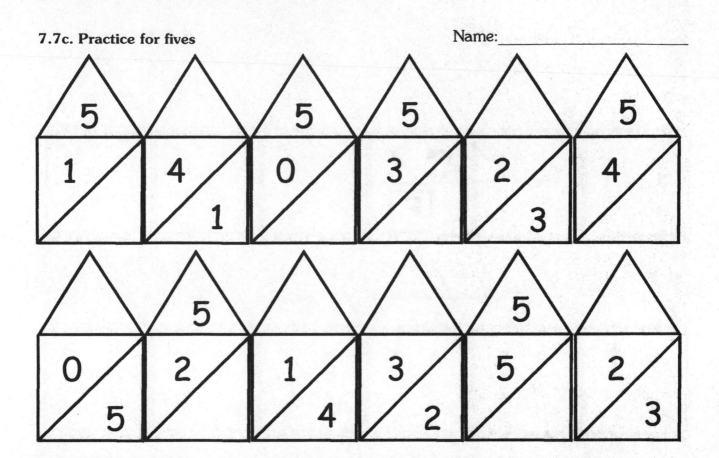

7.7d. Practice for sixes

Name:_____

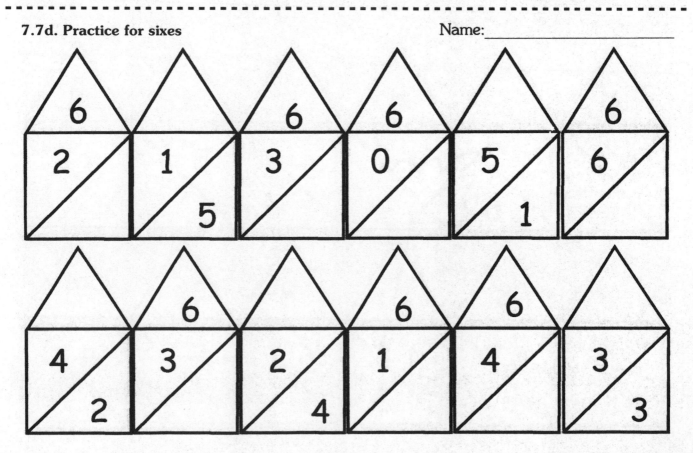

7.7e. Practice for sevens

Name:_____

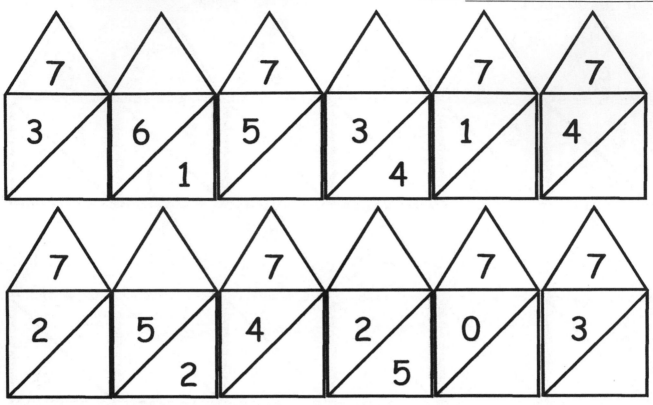

- -

7.7f. Practice for eights

Name:_____

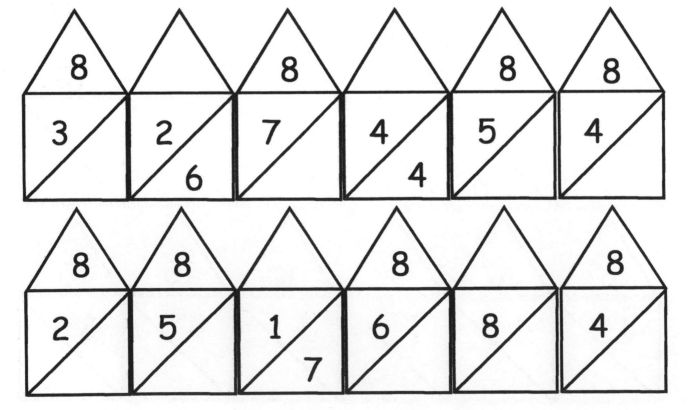

Addition and Subtraction, © 2002 Zephyr Press, Chicago • www.zephyrpress.com

7.7g. Practice for nines   Name:_____

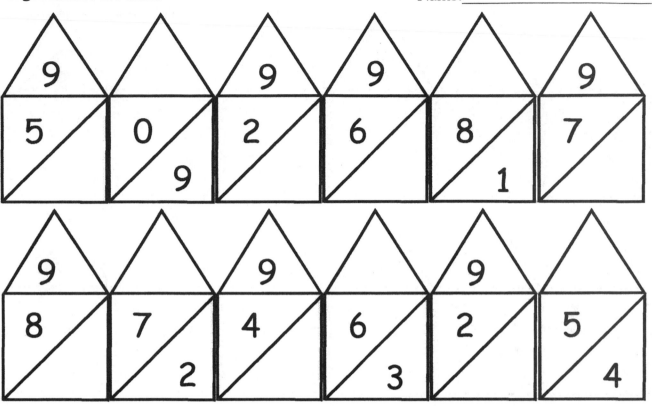

- -

7.7h. Practice for tens   Name:_____

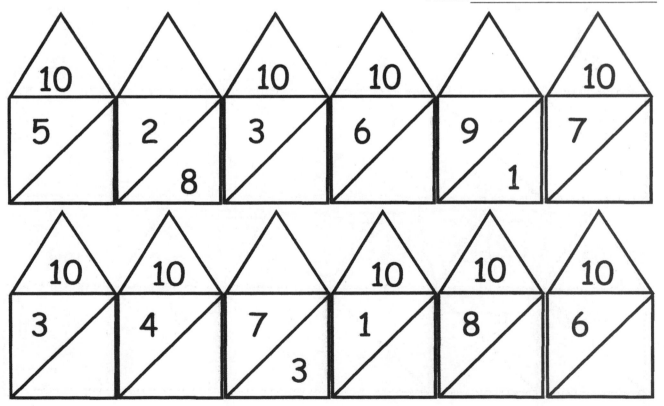

7.8a. Which house? Threes

Name: _____

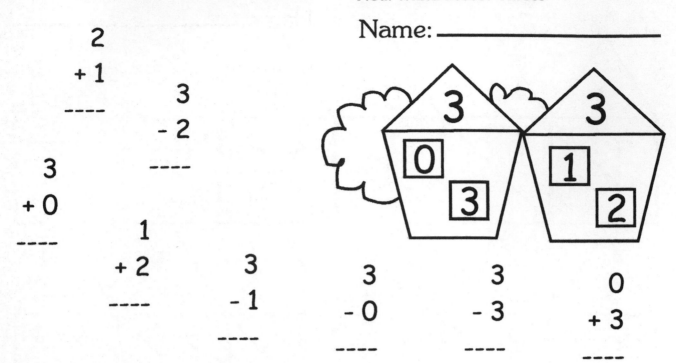

```
  2            3
+ 1          - 2
----         ----

  3            1
+ 0          + 2        3
----         ----     - 1
                      ----
```

```
  3        3        0
- 0      - 3      + 3
----     ----     ----
```

Draw a line from each problem to the house from which it came.

. . . .

7.8b. Which house? Fours

Name: _____

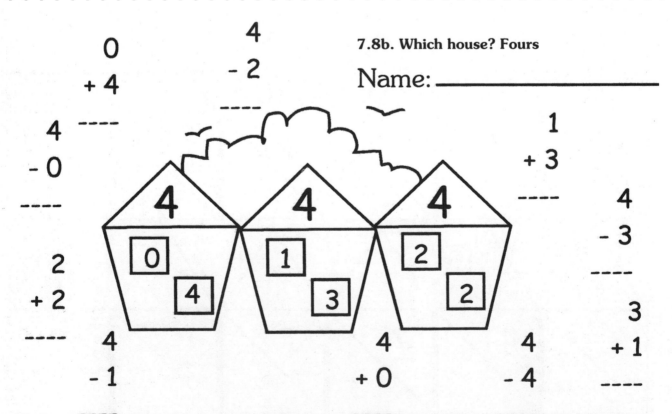

```
  0          4
+ 4        - 2
----       ----

  4                      1
- 0                    + 3
----                   ----

  2                      4
+ 2                    - 3
----                   ----

  4        4        4    3
- 1      + 0      - 4  + 1
----     ----     ----  ----
```

Draw a line from each problem to the house from which it came.

 Addition and Subtraction, © 2002 Zephyr Press, Chicago • www.zephyrpress.com

7.8c. Which house? Fives

Name: _____

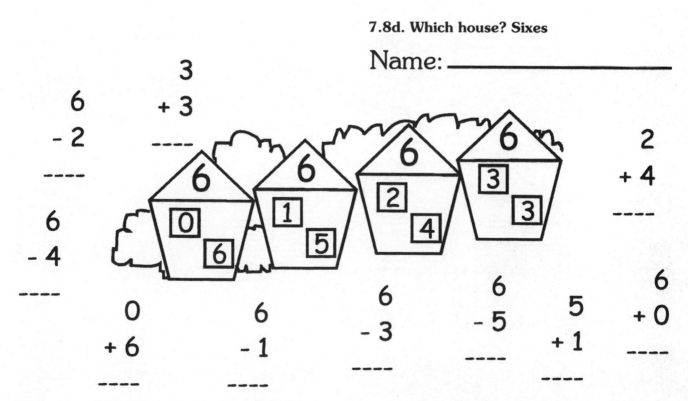

```
  2              5
+ 3            - 3
----           ----

5
+ 0

----     1
       + 4              5           5
       ----           - 2         - 1
                5     ----        ----
       5      - 4     5
     - 4      ----  - 5
     ----           ----     5
                             - 1
                             ----
```

Draw a line from each problem to the house from which it came.

· ·

7.8d. Which house? Sixes

Name: _____

```
  6       3
- 2     + 3                          2
----    ----                       + 4
                                   ----
6
- 4
----     0       6         6     6     6
       + 6     - 1       - 3   - 5   + 0
       ----    ----      ----  5     ----
                               + 1
                               ----
```

Draw a line from each problem to the house from which it came.

7.8e. Which house? Sevens

Draw a line from each problem to the house from which it came.

Name: _____

```
      7
     -5
5    ----
+2
----
```

```
  2
 +5
6 ----
+1
----
```

```
 7
-2
----
```

```
  3
 +4
0 ----
+7
----
```

```
 7
-3
----
```

```
 7
-0
----
```

```
 7
-6
----
```

```
 1
+6
----
```

```
 7
-1
----
```

```
 7
-4
----
```

```
 7
-7
----
```

```
 7
+0
----
```

```
 4
+3
----
```

Addition and Subtraction, © 2002 Zephyr Press, Chicago • www.zephyrpress.com

7.8f. Which house? Eights

Draw a line from each problem
to the house from which it came.

Name: _____

```
        7
       + 1
   8   ----
  - 2              6
  ----           + 2          3
4               ----         + 5      0
+ 4        8                 ----    + 8
----      - 1          8             ----
         ----        - 4
                     ----
8
- 6
----
```

```
        5
       + 3
  8    ----
 - 5                              8
 ----     2          8     1     - 0
         + 6       - 3   + 7     ----
         ----      ----  ----
              8
             - 7
             ----
```

7.8g. Which house? Nines

Draw a line from each problem
to the house from which it came.

Name: _____

$$9 \atop -1 \atop \overline{}$$

$$0 \atop +9 \atop \overline{}$$

$$9 \atop -0 \atop \overline{}$$

$$1 \atop +8 \atop \overline{}$$

$$9 \atop -4 \atop \overline{}$$

$$8 \atop +1 \atop \overline{}$$

$$9 \atop -1 \atop \overline{}$$

$$9 \atop -8 \atop \overline{}$$

$$2 \atop +7 \atop \overline{}$$

$$9 \atop -7 \atop \overline{}$$

$$7 \atop +2 \atop \overline{}$$

$$6 \atop +3 \atop \overline{}$$

$$4 \atop +5 \atop \overline{}$$

$$9 \atop -3 \atop \overline{}$$

$$2 \atop +7 \atop \overline{}$$

$$9 \atop -2 \atop \overline{}$$

$$9 \atop -3 \atop \overline{}$$

Addition and Subtraction, © 2002 Zephyr Press, Chicago • www.zephyrpress.com

7.8h. Which house? Tens

Draw a line from each problem to the house from which it came.

Name: _____

$$10$$
$$- 5$$
$$----$$

$$10$$
$$+ 0$$
$$----$$

$$9$$
$$+ 1$$
$$----$$

$$10$$
$$- 1$$
$$----$$

$$8$$
$$+ 2$$
$$----$$

$$10$$
$$- 4$$
$$----$$

$$10$$
$$- 6$$
$$----$$

$$10$$
$$- 2$$
$$----$$

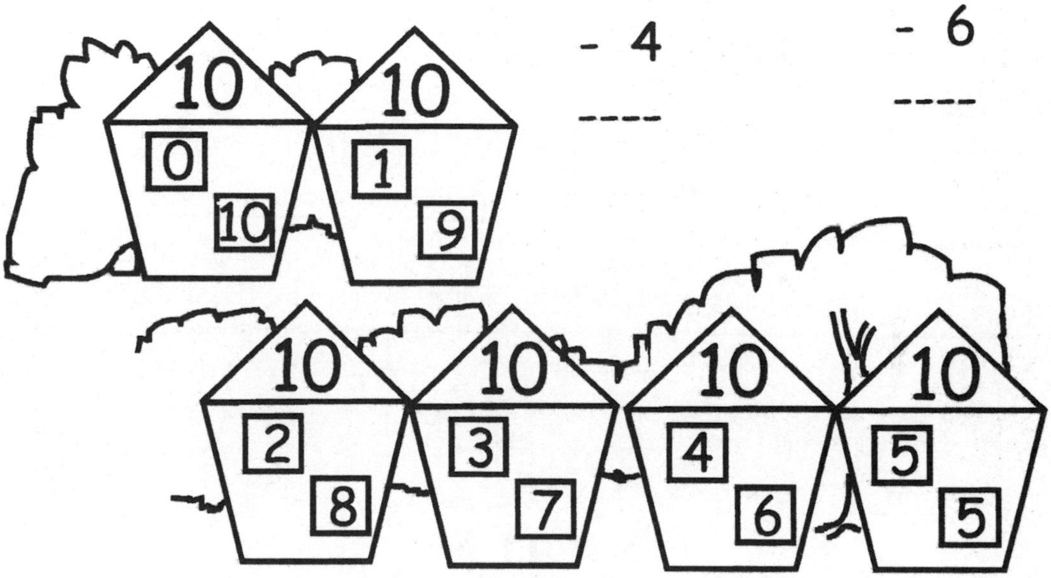

$$10$$
$$- 8$$
$$----$$

$$3$$
$$+ 7$$
$$----$$

$$8$$
$$+ 2$$
$$----$$

$$5$$
$$+ 5$$
$$----$$

$$7$$
$$+ 3$$
$$----$$

$$4$$
$$+ 6$$
$$----$$

$$2$$
$$+ 8$$
$$----$$

7.9a. Writing problems, threes

Write problems
for each house:

Name: _____

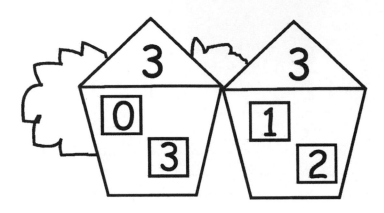

. . . .

7.9b. Writing problems, fours

Write problems
for each house:

Name: _____

Addition and Subtraction, © 2002 Zephyr Press, Chicago • www.zephyrpress.com

7.9c. Writing problems, fives

Write problems
for each house:

Name: _____

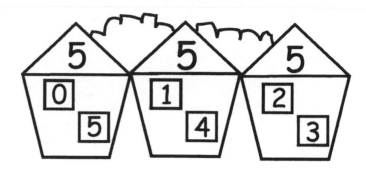

· ·

7.9d. Writing problems, sixes

Write problems
for each house:

Name: _____

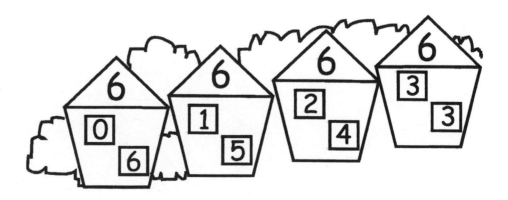

7.9e. Writing problems, sevens

Write problems
for each house:

Name: _____

 Addition and Subtraction, © 2002 Zephyr Press, Chicago • www.zephyrpress.com

7.9f. Writing problems, eights

Write problems
for each house:

Name: _____

7.9g. Writing problems, nines

Write problems
for each house:

Name: _____

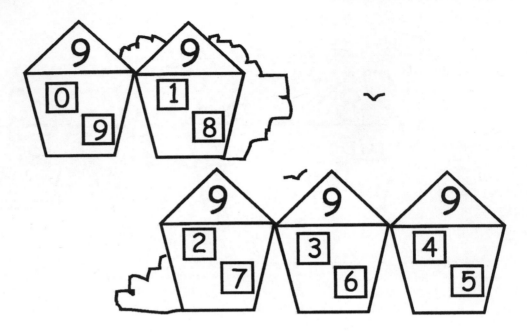

7.9h. Writing problems, tens

Write problems for each house:

Name: _____

7.10a. Addition and subtraction practice, threes and fours

Name: _____

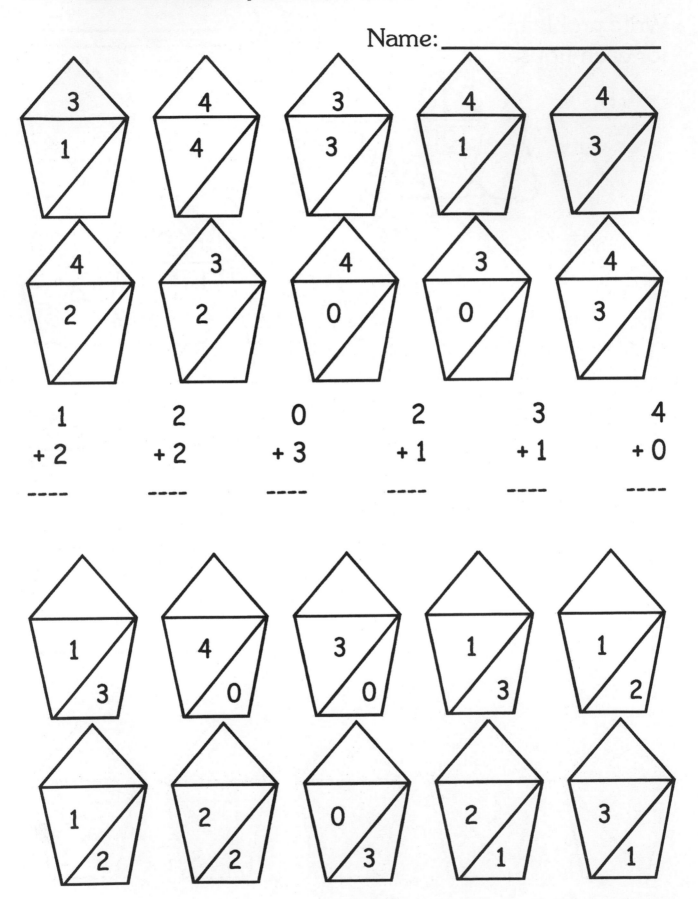

7.10b. Addition and subtraction practice, fives and sixes

Name:_____

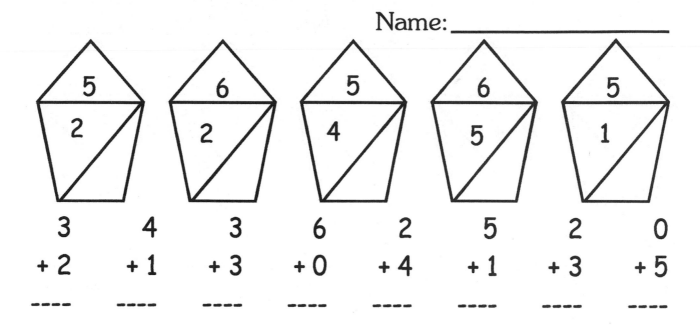

3	4	3	6	2	5	2	0
+2	+1	+3	+0	+4	+1	+3	+5
----	----	----	----	----	----	----	----

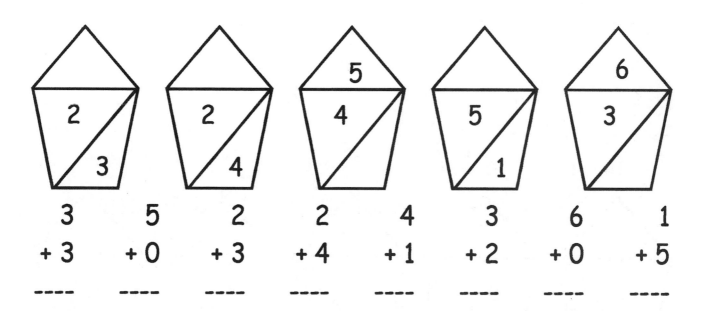

3	5	2	2	4	3	6	1
+3	+0	+3	+4	+1	+2	+0	+5
----	----	----	----	----	----	----	----

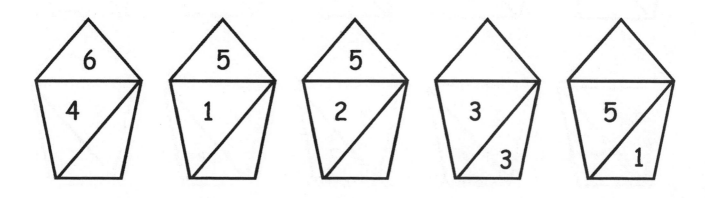

7.10c. Addition and subtraction practice, threes, fours, fives, and sixes

Name: _____

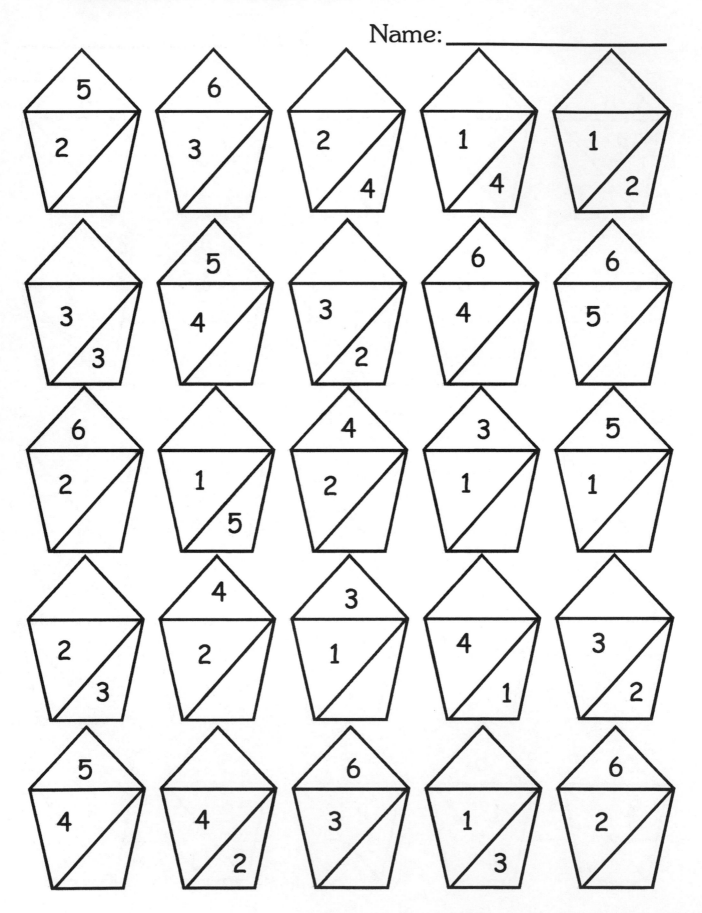

Addition and Subtraction, © 2002 Zephyr Press, Chicago • www.zephyrpress.com

7.10d. Addition and subtraction practice, sevens and eights

Name: _____

5	3	3	7	2	6	4
+ 2	+ 4	+ 5	+ 1	+ 6	+ 1	+ 4
----	----	----	----	----	----	----

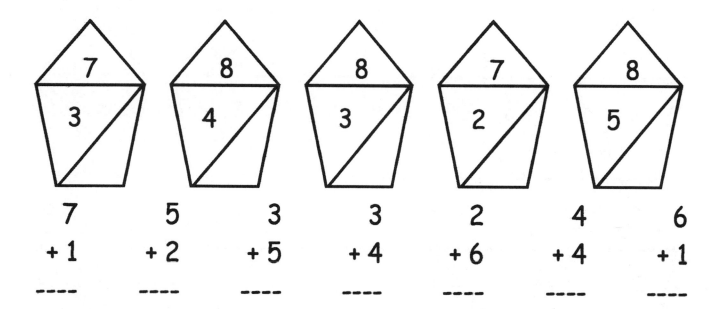

7	5	3	3	2	4	6
+ 1	+ 2	+ 5	+ 4	+ 6	+ 4	+ 1
----	----	----	----	----	----	----

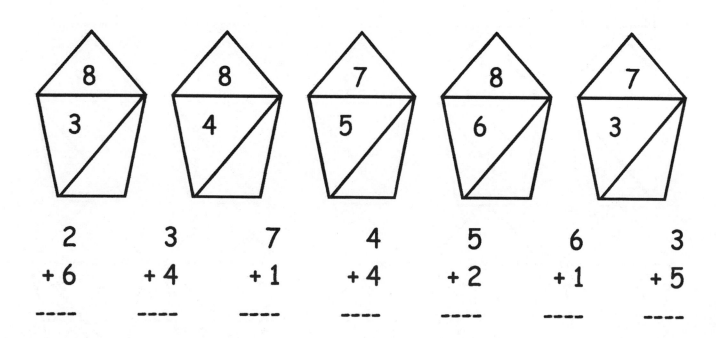

2	3	7	4	5	6	3
+ 6	+ 4	+ 1	+ 4	+ 2	+ 1	+ 5
----	----	----	----	----	----	----

7.10e. Addition and subtraction practice, nines and tens

Name:_____

8	5	5	7	6	6	2
+ 2	+ 4	+ 5	+ 3	+ 3	+ 4	+ 7
----	----	----	----	----	----	----

 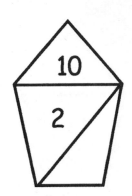 

7	9	4	6	0	5	6
+ 2	+ 1	+ 5	+ 4	+ 9	+ 5	+ 3
----	----	----	----	----	----	----

2	3	5	7	4	6	2
+ 7	+ 7	+ 4	+ 3	+ 6	+ 3	+ 8
----	----	----	----	----	----	----

7.10f. Addition and subtraction practice, sevens, eights, nines, and tens

Name:_____

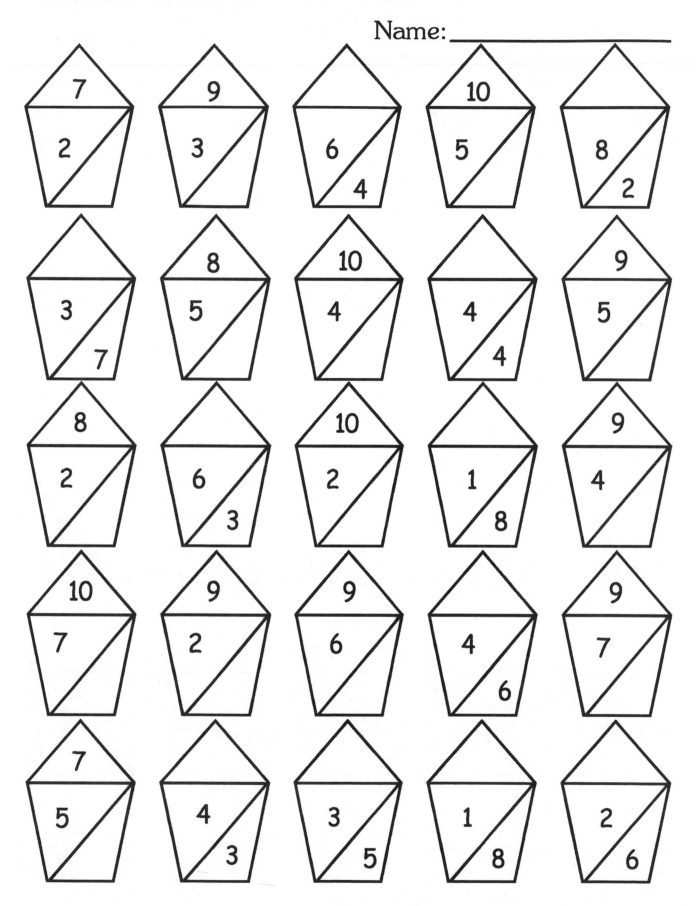

7.11. Extra practice houses

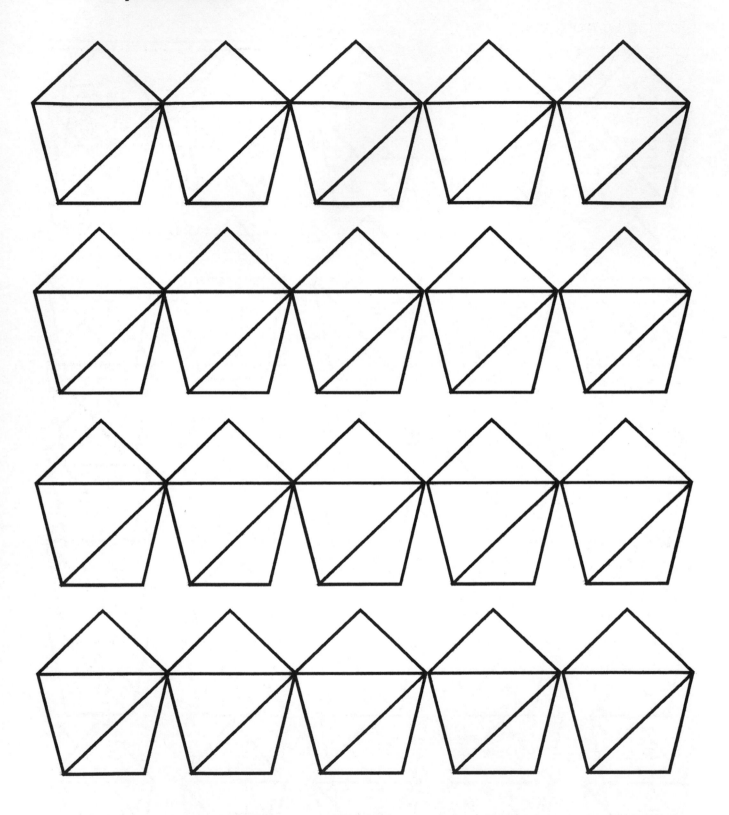

Name_____

Addition and Subtraction, © 2002 Zephyr Press, Chicago • www.zephyrpress.com

7.12a. Mixed subtraction word problem

One family from each house went to a play. You are the bus driver. You must take each family to the right house. Use the numbers in the bus as many times as you need to.

Name: _____

4 6 7 5 3 2

Stony Brook Village

| 10 | 8 | 7 | 9 | 6 |
| 3 | 5 | 5 | 4 | 4 |

| 5 | 9 | 10 | 8 | 7 |
| 3 | 4 | 6 | 6 | 3 |

| 7 | 10 | 8 | 9 | 9 |
| 2 | 5 | 3 | 3 | 2 |

7.12b. Mixed addition word problem

Name: _____

You are the new soccer coach. You need to choose which teams play each other. Place players on the field. Make sure there are the same number of players on each side. The first problem is done for you.

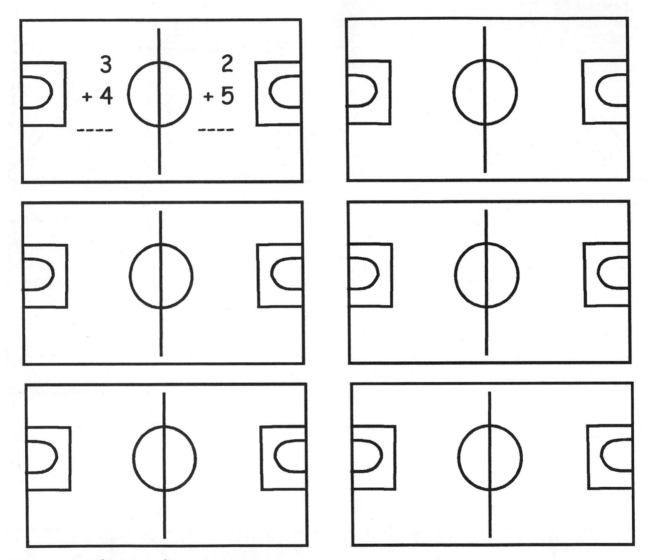

Choose from these teams:

6	2	5	5	6	5	2	7	8	8
+ 4	+ 7	+ 5	+ 4	+ 3	+ 3	+ 6	+ 3	+ 1	+ 2
----	----	----	----	----	----	----	----	----	----

Addition and Subtraction, © 2002 Zephyr Press, Chicago • www.zephyrpress.com

7.12c. Mixed subtraction word problem

Name: _____

It was the big Summer Celebration. But right at the start, a big wind began to blow. Each family at the picnic lost 3 paper plates. How many did each family have left?

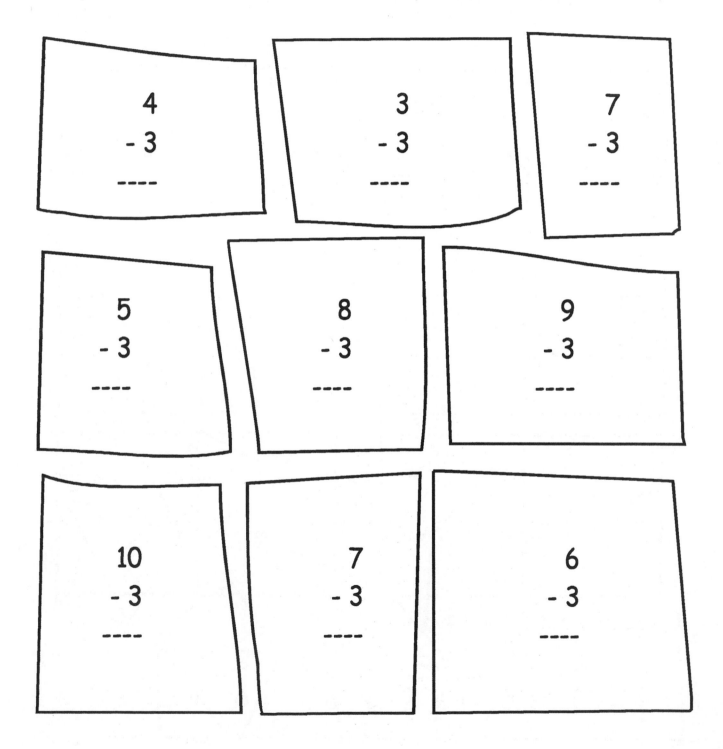

7.12d. Mixed addition word problem

Name:_____

A big wind blew all the attic numbers off! Please help put them back on!

Addition and Subtraction, © 2002 Zephyr Press, Chicago • www.zephyrpress.com

7.12e. Draw your own problems

Name:_____

Choose a street you want to draw. Draw the houses and fill in the families that live in each house. Decorate the street any way you wish.

7.12f. Draw your own problems

Name:_____

Make up a story about any street. Draw numbers in the houses to show what happens in your story.

  Addition and Subtraction, © 2002 Zephyr Press, Chicago • www.zephyrpress.com

7.13a. House flash cards—page 1

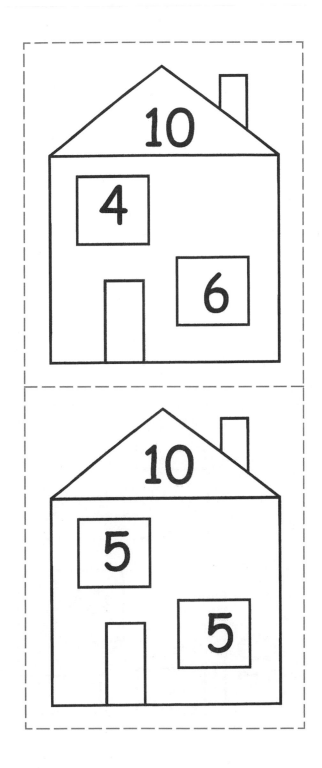

Stony Brook Village Flash Cards

Ideas for Use

Use these flash cards after the child has had plenty of exploration with the felt houses. The cards can be used for review, or a child can refer to them while filling out a practice sheet.

To practice addition facts, cover the attic number. For subtraction facts, cover one window. Two children can pair up to review, with one child covering the numbers and the other giving the answers.

Note

Focus on helping the child discover patterns in the math families. Several patterns are described in the text.

Preparation

Photocopy the flash-card pages. Laminate for durability. Cut apart to make cards. For use as a memory aid, you may wish to punch a hole in one corner and string the set of cards, in order, on a metal ring or length of yarn.

7.13b. House flash cards—page 2

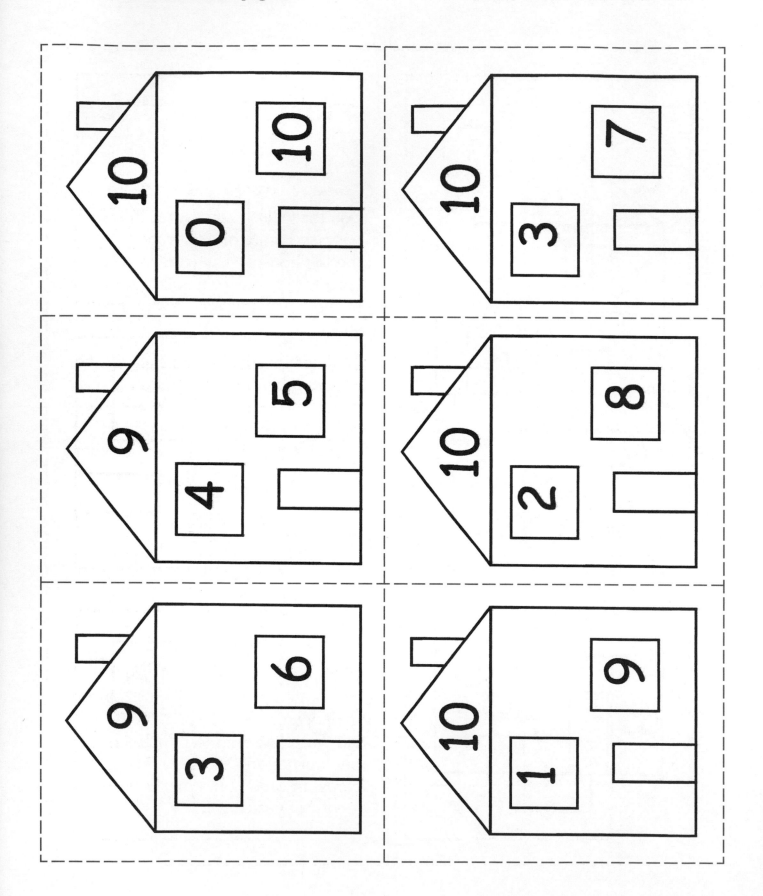

7.13c. House flash cards—page 3

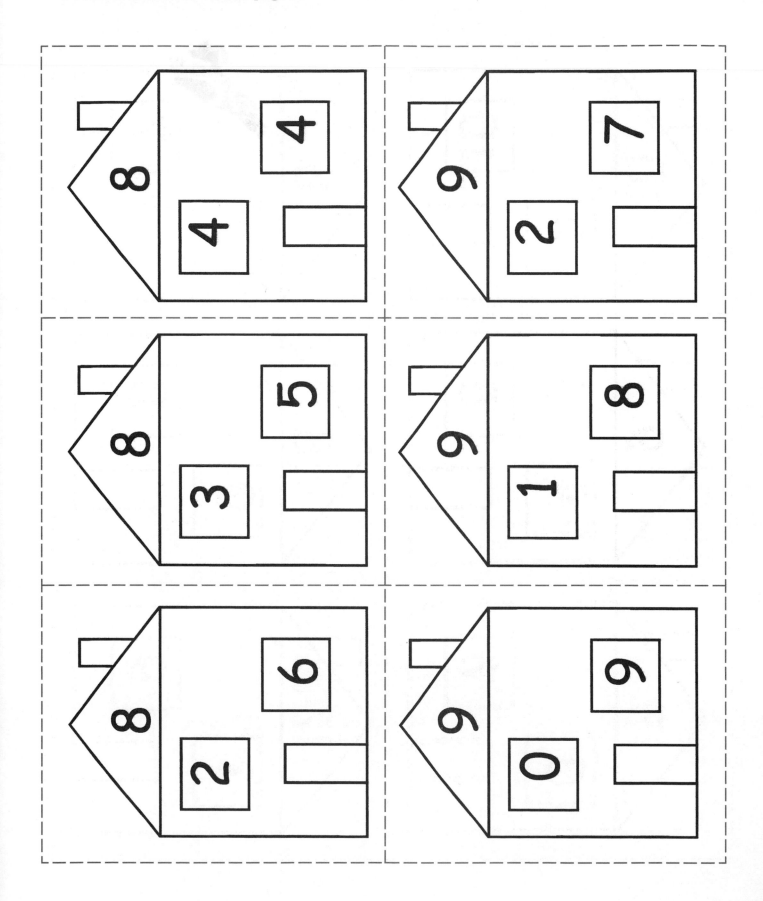

7.13d. House flash cards—page 4

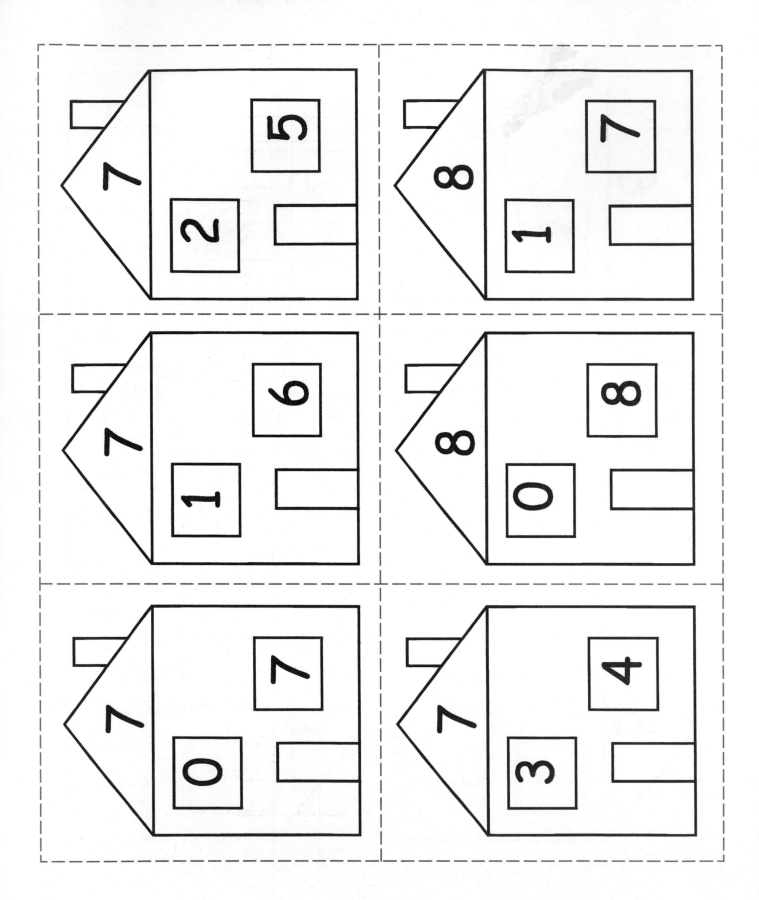

Addition and Subtraction, © 2002 Zephyr Press, Chicago • www.zephyrpress.com

7.13e. House flash cards—page 5

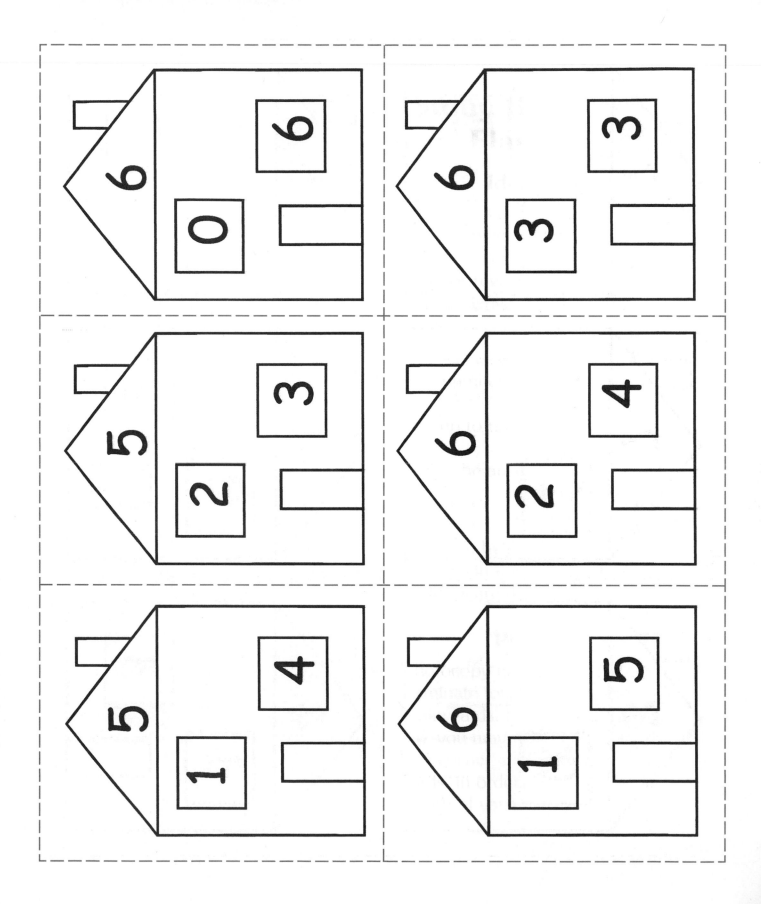

7.13f. House flash cards—page 6

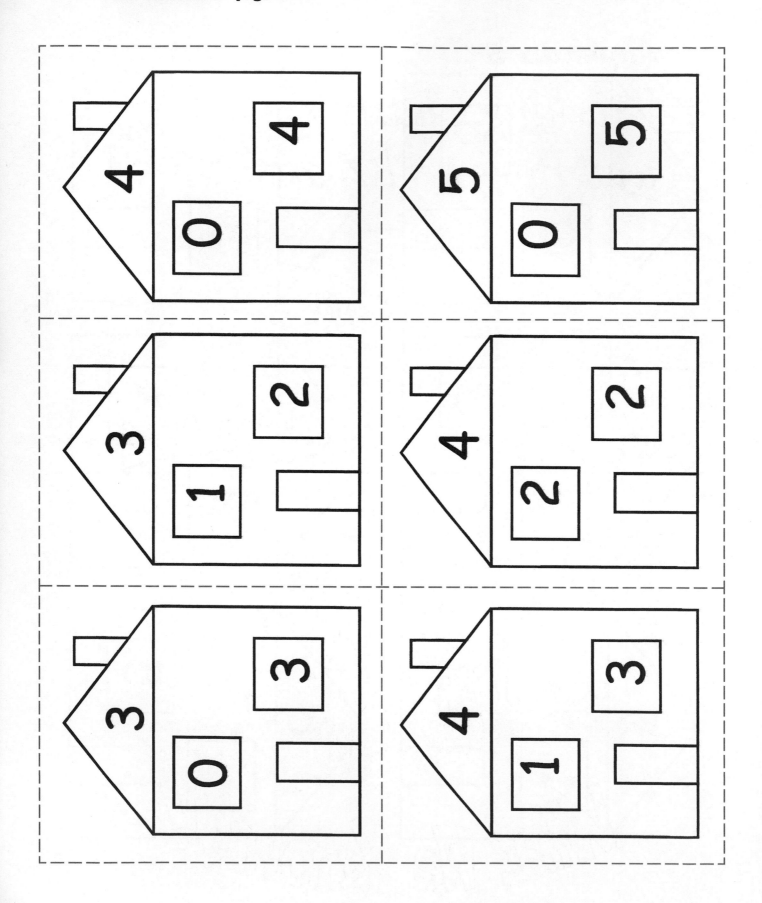

Appendix B
Tracking Charts and Progress Reports

Clipboard Assessment for Learning Numbers (Chapter 4)

Skill:	1		2		3		4		5	6	
	Recognize numbers to ten		Count to ten		Order numbers sequentially		Supply missing numbers		Discover patterns I = interested V = very interested T = taking it further	Write numbers	
Name	1–10	11–20	1–10	11–20	1–10	11–20	1–10	11–20		1–10	11–20

my
numbers

Progress Report (Chapter 4)

_____ 's Numbers   Date: _____

	10	20	Other
1. I can recognize numbers to			
2. I can count to			
3. I can order numbers to			
4. I can supply missing numbers to.....			
5. I can write numbers to			

Teacher's comments:

- -

my
numbers

Progress Report (Chapter 4)

_____ 's Numbers   Date: _____

	10	20	Other
1. I can recognize numbers to			
2. I can count to			
3. I can order numbers to			
4. I can supply missing numbers to.....			
5. I can write numbers to			

Teacher's comments:

Whole Class Record (Chapters 5–7)

Check off or date each skill when student shows mastery

Name	Dot Cards	Fives Frames to 20	Fives Frames to 70	My Two Hands	Houses

Addition and Subtraction, © 2002 Zephyr Press, Chicago • www.zephyrpress.com

Whole Class Record (Chapter 7)

Check off or date each street when student shows mastery

Name	Third	Fourth	Fifth	Sixth	Seventh	Eighth	Ninth	Tenth

Individual Assessment: Computation to Ten (Chapter 7)

Name _____

Based on several observations and samples of written work, draw a line through each problem as it is mastered. When an entire street has been mastered, enter the date of mastery. Note any needs for additional practice in the last column.

Street						Date of Mastery	Extra Practice Needed
Third:	0 + 3	1 + 2					
Fourth:	0 + 4	1 + 3	2 + 2				
Fifth:	0 + 5	1 + 4	2 + 3				
Sixth:	0 + 6	1 + 5	2 + 4	3 + 3			
Seventh:	0 + 7	1 + 6	2 + 5	3 + 4			
Eighth:	0 + 8	1 + 7	2 + 6	3 + 5	4 + 4		
Ninth:	0 + 9	1 + 8	2 + 7	3 + 6	4 + 5		
Tenth:	0 + 10	1 + 9	2 + 8	3 + 7	4 + 6 5 + 5		

- -

Individual Assessment: Computation to Ten (Chapter 7)

Name _____

Based on several observations and samples of written work, draw a line through each problem as it is mastered. When an entire street has been mastered, enter the date of mastery. Note any needs for additional practice in the last column.

Street						Date of Mastery	Extra Practice Needed
Third:	0 + 3	1 + 2					
Fourth:	0 + 4	1 + 3	2 + 2				
Fifth:	0 + 5	1 + 4	2 + 3				
Sixth:	0 + 6	1 + 5	2 + 4	3 + 3			
Seventh:	0 + 7	1 + 6	2 + 5	3 + 4			
Eighth:	0 + 8	1 + 7	2 + 6	3 + 5	4 + 4		
Ninth:	0 + 9	1 + 8	2 + 7	3 + 6	4 + 5		
Tenth:	0 + 10	1 + 9	2 + 8	3 + 7	4 + 6 5 + 5		

  Addition and Subtraction, © 2002 Zephyr Press, Chicago • www.zephyrpress.com

References

Barbe, Walter B. 1985. *Growing Up Learning*. Washington, D.C.: Acropolis Books.

Caine, Geoffrey, Renate Nummela Caine, and Sam Crowell. 1999. *MindShifts: A Brain-Compatible Process for Professional Growth and the Renewal of Education*. Rev. ed. Tucson, Ariz.: Zephyr Press.

Gardner, Howard. 1993. *Frames of Mind: The Theory of Multiple Intelligences*. 2nd. ed. New York: Harper and Row.

Gregorc, Anthony F. 1982. *An Adult's Guide to Style*. Columbia, Conn.: Gregorc Associates. (Available from 5 Doubleday Rd., Columbia, Conn. 06237. (203) 228-0093.)

Hart, Leslie. 1999. *Human Brain and Human Learning*. Kent, Wash.: Books for Educators.

Jensen, Eric. 1994. *The Learning Brain*. Del Mar, Calif.: Turning Point.

————. 1998. *Teaching with the Brain in Mind*. Alexandria, Va.: ASCD.

LeDoux, Joseph. 1996. *The Emotional Brain: The Mysterious Underpinnings of Emotional Life*. New York: Simon and Schuster.

Swassing, Raymond, and Walter Barbe. 1999. Swassing-Barbe Modality Index. In *Modality Kit*. Columbus, OH: Zaner-Bloser, Inc.

Tobias, Cynthia Ulrich. 1994. *The Way They Learn*. Colorado Springs, Colo.: Focus on the Family Publishing.

Witkin, Herman A. 1977. Cognitive Styles in the Educational Setting. *New York University Education Quarterly*, 14–20.

Index

Note: Page numbers in **bold** refer to Reproducible Blackline Masters

About the Author

Sarah Morgan Major does extensive Title 1 teaching with grades PreK through third in a public school. She is a fluent Spanish speaker and is the primary ESOL (English for Speakers of Other Languages) instructor in her school where she is able to use the innovative reading program she has developed. In addition, Sarah provides training, materials, and support to classroom teachers to help them meet learners' needs within the regular classroom. She holds a B.A. in art from Wheaton College, Wheaton, Illinois; and an M.Ed. from Aquinas College in Grand Rapids, Michigan. Her passion and  expertise are in providing learning tools and teacher support in order to enable struggling students to excel in their regular classrooms. During her years of working with students who struggle in math and reading, Sarah developed a reading method that is now being used within three Michigan school districts in regular and special education classrooms. She has also developed a three-book series, *Kid-Friendly Computation,* that progresses from learning numbers to multiplication and division. The materials and methods are rich in visual and kinesthetic components that make learning quick and effortless for the students. Sarah's passions outside of work include her two artist-musician children, Matthew and Melissa; gardening; beaching it; music; books; and writing.